# CLOUDS OF THE WORLD

A Complete Colour Encyclopedia

# CLOUDS OF THE WORLD

A Complete Colour Encyclopedia

***RICHARD SCORER***

LOTHIAN PUBLISHING CO (PTY) LTD · MELBOURNE

DAVID & CHARLES · NEWTON ABBOT

ISBN 0 7153 5485 X (Great Britain)
ISBN 85091 1168 (Australia)



First published 1972



Library of Congress Cataloging in Publication Data

Scorer, Richard Segar, 1919-
Clouds of the world.

Bibliography: p.
1. Clouds—Pictorial works. I. Title.
QC921.S357 551.5'76 72-1115
ISBN 0-8117-1961-8

Platemaking by Lawrence-Allen Ltd
Weston-super-Mare
Printed in Great Britain by Gildenprint Ltd
Bristol

Published in Great Britain by
David & Charles (Publishers) Ltd
Newton Abbot Devon
and in Australia by
Lothian Publishing Co Ltd
Melbourne

# CONTENTS

# PREFACE

The industrial revolution followed by today's fashionable philosophy of maximised economic growth have killed man's awe of Nature, replacing it with a contemptuous view of the Earth as a source of wealth to be exploited without shame. But the Earth is our home, and it is a delight to stand back in wonder at its beauty.

This planet stands alone in the solar system as a host for life. It is unique in its variety of geographical patterns and in the liveliness of its changing skies, and their continual evolution and the flux of air and water have been endowed in literature with a spiritual existence. Nevertheless they are mere physical and mechanical systems, and to understand them as such is to exorcise the spirits. Then can a man love the experience of storms as much as of colourful skies, and can feel for the subtle shapes of clouds without false anthropomorphisms. Nor does he need the device of poetic mystery; for knowledge brings with it not only a heightened sense of awe but a deep awareness of our total ignorance which is too real to be made light of by trite metaphors.

So: the clouds are arranged according to the physical and mechanical evolutions through which they quickly pass: they are always changing. Those who know no meteorology can search these pages visually for familiar scenes, and will soon find out the basis of the arrangement. Those with greater pretensions will find their way around the index and list of plates. Some may be surprised at the absence of, or minor role taken by well known names; but will discover instead words even more familiar and just as useful.

Thus cumulonimbus takes a back seat in favour of showers, squalls, and cold fronts. Lenticular clouds are called wave clouds when that is what they are, and something else when it is not. Mist is merely thin fog unless the fog is thick haze. Cirrocumulus does not really exist at all: insofar as it does so officially it is merely a kind of altocumulus which differs from stratocumulus not in height or apparent size as the handbooks would have believed, but in being a layer that is cellular rather than an upcurrent that has spread out. Billows are clearly distinguished from altocumulus, and from waves, with which they are often confused. Nimbostratus is not illustrated: it is a grey amorphous raining sky, and who wants to get a good camera wet to photograph that!

We all tend to have our own favourite pictures which are preservations of exciting but often fleeting moments when the desire to gaze competes with attending to exposure meter, lenses, film, and notebook. This is followed by anxious days before the film is finally developed, and many disappointments go with the few real successes because the eye is so much more discriminating than the camera. Among my own favourites are 1.1.5, 1.3.5, 1.5.3, 5.7.3, 9.4.1, 11.1.2 and 13.2.6. Among pictures supplied by others I can recommend the following for pure drama: 2.2.6, 3.3.4, 3.6.1, 5.1.3, 5.4.5, 5.7.9, 9.2.7, 11.2.1, 13.3.3, 13.4.4, 14.3.8 and 14.4.8. I wish I had been there.

Beautiful though these pictures are, it is through simplified understanding that we come to like them. For this reason diagrams have been used extensively to explain the motion.

Through the generous cooperation of many observers who are named below it has been possible to make this collection nearly worldwide, far beyond the extent of my own travels. Actually clouds are not very different the world over; it is just that each type has areas where it is most common and best exemplified. Besides this global view, and the use made of several satellite pictures, we have introduced stereoviews. They are displayed so that those who can learn the cross-eyed technique can see them in 3D without optical aids. Not everyone will be able to do this and so we have provided some duplicate prints which can be detached and used for conventional stereoscopic viewing (eg with a 10cm lens to each eye).

Clouds are not biological species which reproduce themselves almost identically. They are all different, and an infinite and continuous variation is possible. Therefore no encyclopaedia can be exhaustive, and it is hoped that through use of these pages readers will become discriminating enough to discern distinctly new forms in the skies they see. Since skies are the same for professional and amateur, the pictures represent familiar things in spite of the technicalities. For this reason they can be of universal interest.

To some the awful beauty of Nature is evidence of a creator. To me it is evidence that Nature is awfully beautiful, which to know is a priceless treasure, an aspect of our environment to be deeply respected, from which we must stand back in wonder and hesitate to spoil. All is not simple cold physics and mechanics because through evolution since life began life has made the atmosphere what is is, recycling it continuously, and has evolved to live in harmony with it as it has taken new shape. The very fact that our atmosphere is like it is would be evidence to a visitor from outer space not merely that life could be here but that it must be here.

In the hope that man may quickly come to respect this mutual dependence and learn to enjoy, not grab and deface, what is in the world, I offer this record of its grandeur.

R. S. Scorer
October 1971

# ACKNOWLEDGMENTS

Special acknowledgement is due to a few friends who have by their spontaneity enlarged my whole view of the subject: above all, Frank Ludlam for whom clouds have been a fascination since boyhood. With Betsy Woodward (now Proudfit), Harold Klieforth, Milan Koldovsky, David Pedgley, Harry Wexler, and Alistair Fraser, I have had the privilege of long discussions, and I have been inspired by Tetsuya Fujita's superb professionalism in analysing storm pictures. Dr Tricker's explanation of the Lowitz arc seems to be the right one at last; Professor J L Lumley told me how to use the cross-eyed method of stereo-viewing, and Les Hubert introduced me to many new aspects of satellite pictures.

For other pictures acknowledgement is gratefully made to Roger Akester, Blake Allison, Toshio Aono, The Canadian Department of Transport, Brian Carroll, D M Cookson (and his father and brother), G R Cresswell, Robert M Cunningham, Joan Darby, Miss B Evans, J H Feige, John Findlater, Ben Fogle, J Frizzola (who got M Rosen to take 13.3.3 at the Brookhaven National Laboratory), Michael Garrod, D G Harper and the Controller of H M Stationery Office, Eigil Hesstvedt (for the Mother of Pearl Clouds by A Holt and H G Hemm), J J Hicks, Steve Hodge, Charles Hosler, N D P Hughes, Roger Jensen, Miss P M King, L Larssen, Peter Lester, Jim Lovill, L A Milner, M Martinelli, Elizabeth Mayo, James H Meyer, P G Mott (Hunting Aerosurveys), N.A.S.A. (for the satellite pictures), John Nielan, M Pesout, Vernon G Plank, J H Reuss (and the Technische Hochschule, Darmstadt), C Rey, Claude Ronne (and the Woods Hole Oceanographic Institution), The Royal Air Force, H H Samuelson, Vincent Schaefer, Emil Schulthess, Beatrice Scorer, Joanne Malkus Simpson, Bob Simpson, S Suzuki, Robert Symons, P F Taylor, G Thompson, S A Thorpe, C T Tilbrook, N Tinbergen, Carol Unkenholz, Kenneth Williamson, Philip Wills, Gerhard Vogt, the Stormy Weather Group of McGill University, and the U S Weather Bureau (as it was when R H Simpson took his magnificent hurricane pictures).

# LIST OF PLATES

# HISTORY OF CLOUD CLASSIFICATIONS

*by* F. H. Ludlam

Clouds have an infinite variety of *shapes*, but a limited number of *forms* corresponding to different physical processes in the atmosphere which are responsible for their formation and evolution. The first careful classifications of clouds were made at the beginning of the 19th century, at a time when none of these processes was clearly understood. Nevertheless the French naturalist Lamarck in 1802 and the English pharmacist Luke Howard in the following year both recognised in their publications the three fundamental forms and some others less basic. The two principal processes responsible for cloud formation are *local convection*, producing scattered individual clouds which usually are domed above a flat base (the *heap*, or cauliflower clouds), and *large-scale convection*, in which air over vast regions ascends at a very shallow inclination to the horizontal and produces the widespread *sheet* clouds which are found in and around bad-weather disturbances such as cyclones. Lamarck called these respectively "nuages groupés" (grouped or piled clouds) and "nuages en voile" (veil or sheet clouds), while Luke Howard used the Latin names "cumulus" (a heap) and "stratus" (a layer). Detached high clouds have a very distinctive fibrous texture and form a third important class which Lamarck called "nuages en balayures" (sweep clouds) and Howard "cirrus" (hair-like, or fibrous). Lamarck distinguished other forms: "nuages en barres" (barred or banded clouds), "nuages pommelées" (dappled or curdled clouds) and "nuages moutonnées" (clouds in flocks). Later he named still more, for example, "nuages en lambeaux" (torn or ragged clouds), and introduced descriptive adjectives such as isolated, obscure, and undulatory. In spite of its soundness and detail Lamarck's classification aroused little notice, even in France, whereas Howard's (perhaps because of his use of Latin names) was successful and has survived to the present day.

Howard considered the other significant cloud forms to be derived from the three basic ones by transition or association. Accordingly they were described by compound names: cirrocumulus, cirrostratus, cumulostratus and cirro-cumulo-stratus or nimbus (the multi-layer rain cloud). The first two were applied to tenuous high clouds which were dappled or spread in a sheet. The third, cumulostratus, is the thunder-cloud: Howard noted that when cumulus builds rapidly a cirrostratus is frequently seen to form around the summit, and then quickly to become denser and spread until a large cloud is formed which may be compared with a mushroom with a very short thick stem.

In 1836 the German meteorologist Kaemtz published a textbook in which he used Howard's classification and added stratocumulus, a lumpy low-level layer cloud. Renou, the Director of French meteorological observatories, also extended the classification soon after 1855 by defining altocumulus and altostratus, dappled and extensive sheet clouds at levels intermediate between those occupied respectively by cirrocumulus (a high cloud) and stratocumulus (a low cloud), and by cirrostratus (a high cloud) and stratus (a cloud always very near the ground). In this way he began the idea of the height of a cloud as a basis for its classification. This idea has a sound physical basis, because the fibrous clouds (cirrus forms) are practically always ice clouds, formed in the high troposphere, while the cumulus, stratus and stratocumulus are formed at low levels by air motions which originate at the ground (for example, in convection provoked by sunshine warming the ground). On the other hand, the large-scale convection produces sheet clouds at all levels in the troposphere, and they often appear in the middle or high troposphere far from regions over which they formed: they are independent of any local air motions near the ground. Consequently clouds can be divided into three categories according to characteristic level of appearance:

1. Low clouds.
2. Intermediate level (alto-) clouds, sometimes frozen.
3. High, frozen clouds: cirrus.

The idea of the importance of height developed, partly because this as yet unrecognised physical basis produced a definite impression of grouping in the observed clouds, and partly because of a growing interest in the general circulation of the atmosphere. At a time when there were no other techniques for observing the currents of air above the ground, great importance fell upon measurements of cloud drift, which could be interpreted as observations of upper-level winds if the cloud heights were known. Since the determination of cloud height from the ground is a difficult and tedious exercise beyond the powers of most observers, it was hoped that the form of the cloud alone might be a sufficiently accurate indication of its level.

The work of observing the forms and heights of clouds became concentrated at a few meteorological observatories, amongst which the most famous was perhaps that at Uppsala in Sweden, directed by Hildebrandsson. In 1880 he published the cloud classification used there, which was Howard's with the addition of stratocumulus and altocumulus. About this time other students were publishing their own classifications, in particular Poëy (1865) and Weilbach (1880) in France, Mühry (1874) in Austria, and the Reverend Clement Ley, of Ashby Parva, in England (1879). Some of these classifications were very detailed and contained significant ideas. Poëy invented the prefix 'fracto' for ragged clouds, Ley introduced the adjectives 'lenticularis' (for the lens-shaped or hump-backed clouds produced when air flows over mountains) and 'castellatus' (for turretted clouds formed when local convection develops in a middle-level layer cloud), and Weilbach substituted the name 'cumulonimbus' for cumulostratus. All these names are still used, but in so far as any writer departed materially from Howard's general scheme he met with little response from other students.

In 1887 the famous Victorian meteorologist the Hon. Ralph Abercromby (who as an enthusiastic amateur made two voyages round the world to assure himself that the same basic cloud forms were to be found everywhere) reviewed the first comprehensive series of cloud height measurements, which had been made at Uppsala. He claimed that they proved a decided tendency for clouds to form at three definite levels: 'it would be premature to speculate on the physical significance of this fact, but we find the same definite layers in the tropics, and no future nomenclature will be satisfactory which does not take the idea of these levels into account.' In the same year he published with Hildebrandsson a classification containing ten principal forms of clouds:

| | | |
|---|---|---|
| High clouds | : | cirrus, cirrostratus |
| Middle clouds | : | cirrocumulus, altocumulus, altostratus |
| Low clouds | : | stratocumulus, nimbostratus |
| Clouds of ascending currents | : | cumulus, cumulonimbus |
| Elevated fog | : | stratus |

In 1889 Clayton, of the Blue Hill Observatory in Massachusetts, published his own cloud height measurements, which did not reveal any tendency for clouds to be especially prevalent at a particular middle level, but there was evidently a strong intuitive belief in the validity of the classification which could not be shaken. In 1891 an International Meteorological Conference endorsed the classification and in 1896 a cloud atlas of 28 coloured plates was published to illustrate it. Since that time the classification has been officially adopted by Meteorological Services everywhere and has evidently proved satisfactory, for only minor amendments have been made on the occasion of the publication of greatly enlarged atlases, first in 1932 and then more recently in 1956. The last issue contains a more elaborate classification than any previously published, in which the ten principal cloud forms are ranked under the heading *genera*, each of which may be sub-divided into *species* and *varieties* and which may possess *complementary features*. Few students of clouds will feel satisfied that this elaboration serves any good purpose, and it can be anticipated that the continuation of the active research into cloud physics which has sprung up in the last two decades will bring a desire to modify the classification in the light of better knowledge of the physical processes at work in clouds. Already, for example, the relegation of the important class of wave clouds to the status of a species seems unfortunate. The preservation of the essentials of the present classification since 1887 should perhaps be seen as a consequence of the paucity of penetrative studies until very recently, but it must be conceded that it has worked well in its prime function of permitting unskilled observers to arrive at a significant description for transmission to a weather analyst, from the appearance of the cloud alone.

# 1 CUMULUS

## 1.1 Thermals and Small Cumulus

Cumulus are clouds produced by thermal convection. They are usually clouds which are evaporating as rapidly as they are being formed. Each cloud has a life of a few minutes and new ones grow as old ones vanish so that the scene looks generally the same for a much longer period.

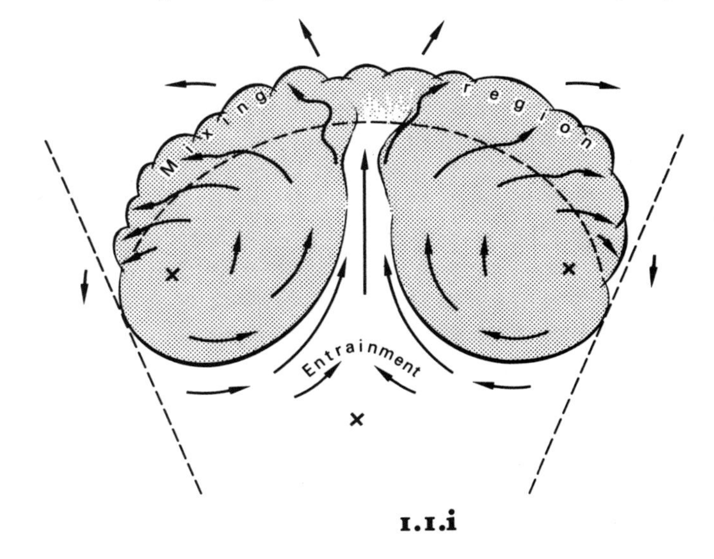


1.1.i

**1.1.i.** shows the configuration of motion in a mass of rising buoyant air, called a thermal. It grows (a) by mixing over its upper surface with the air into which it is advancing, and in a visible thermal at the top of a cumulus cloud this mixing occurs in the cauliflower-like region which has a sharp outline (see 1.1.8), and (b) by the entrainment into its base of exterior air which then rises up the middle where the strongest upcurrent is, and enters the mixing region from below.

The arrows show the relative strengths of the velocities in the circulation and the thermal as a whole rises at somewhat less than half the speed of the upcurrent in the centre.

On account of the entrainment the thermal increases its size along a cone, indicated by the dotted lines, of semi-vertical angle about 15°. The thermal turns itself inside out, the mixing therefore affecting all parts of it, as it rises about $1\frac{1}{2}$ diameters.

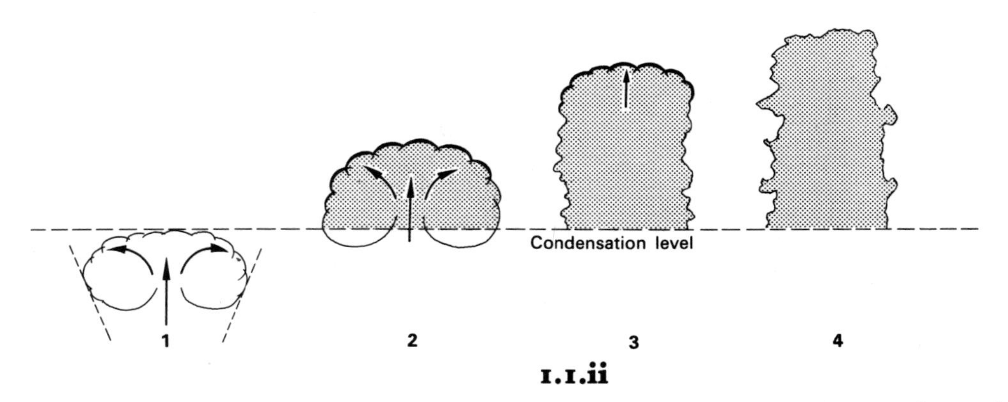


1.1.ii

**1.1.ii.** shows how a thermal often ceases to widen when it ascends above the condensation level and becomes visible as cloud. The surroundings are now stably stratified, which means that diluted parts of the cloud which have less buoyancy are left behind at a lower level. The thermal probably entrains very little air into its base because of the stratification, so that most entrainment must occur at the top by mixing with the air into which it advances.

The condensation of liquid water releases heat which maintains the buoyancy in the interior of the thermal during the ascent, but evaporation on the outside extracts the heat again and causes downdrafts. In the figure the dilute cloud left behind the rising thermal constitutes a cumulus tower. It ceases to grow when the dilution affects the whole thermal, and the top then loses its sharp outline.

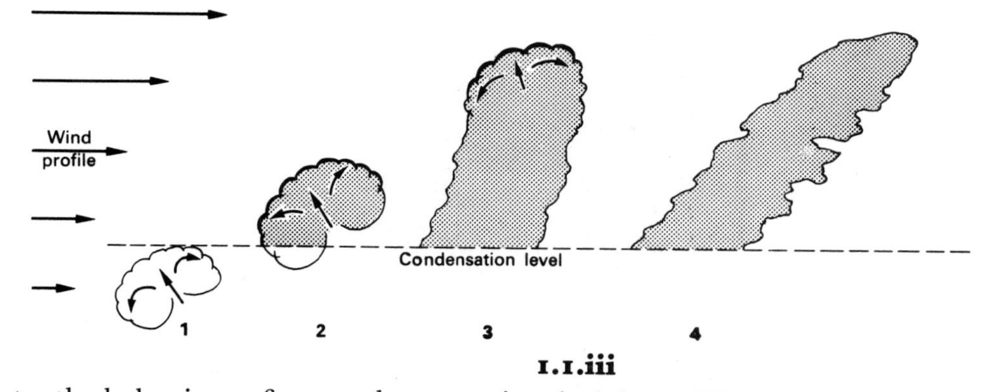


**1.1.iii**

**1.1.iii.** indicates the behaviour of a cumulus tower in wind shear. The axis of the thermal is tilted into the wind but the tower leans over more as time proceeds until it becomes so stretched out that it is quickly evaporated. Because of the distortion and more rapid evaporation cumulus do not grow as large in wind shear: there are no compact moistened regions left by previous thermals into which later ones can rise and evaporate more slowly.

Small cumulus in anticyclonic weather are often seen in stage 2, and the rotation of the downwind (upper) half is usually much more evident than that of the upwind half.

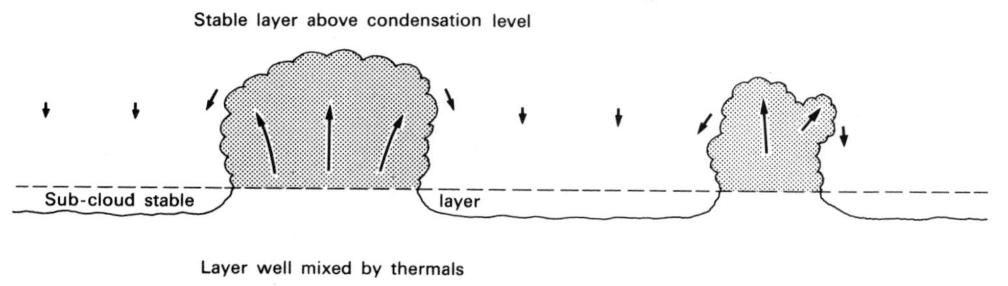


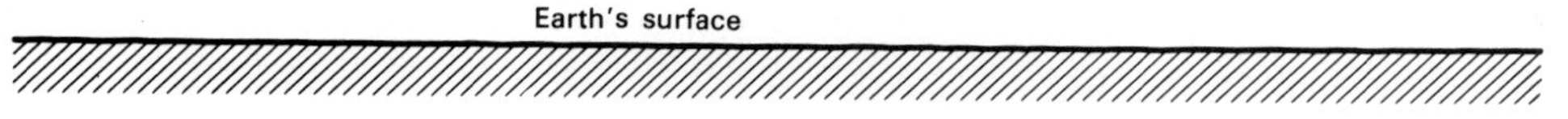


**1.1.iv**

**1.1.iv.** displays the relative magnitudes of typical upcurrents in small cumulus and the sinking motion in between. If this sinking motion occurs, stable air is carried down below the condensation level in between the clouds, and this often stops thermals rising up to the condensation level. There is therefore a tendency for thermals following their predecessors through holes in the *sub-cloud stable layer* to grow preferentially into clouds.

The layer below the stable air is well mixed by thermals from the warm ground and usually has a fairly well marked haze top. This top occurs above the cloud base on occasions when there is a general horizontal convergence which stops the slow sinking motion altogether. Such convergence occurs as a result of sea breezes into small land areas: but it does not usually occur in cold air masses over the ocean unless the clouds are very large and producing substantial amounts of rain.

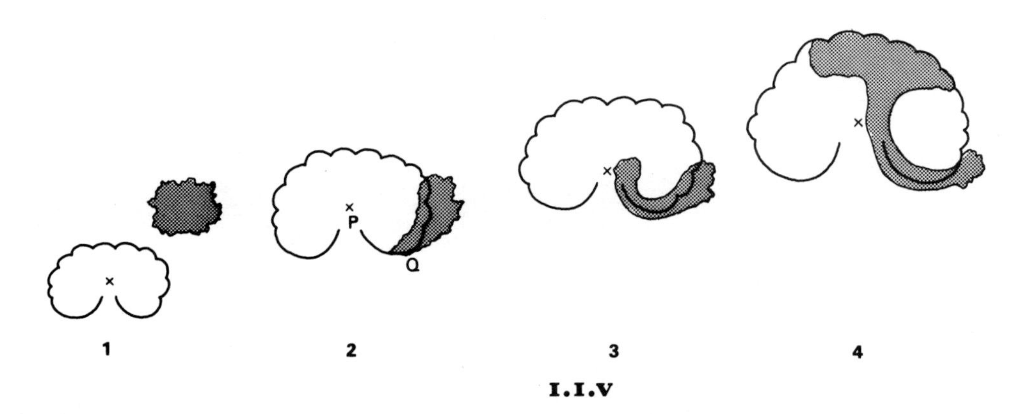


**1.1.v**

**1.1.v.** demonstrates what happens to a small mass of air which lies in the track of a rising thermal. The small mass could be a previous thermal or a glaciated piece of cloud: in this case it is clear that the ice particles would be carried to all parts of the thermal as it rises about $1\frac{1}{2}$ diameters.

The incorporated air is first distorted by the shear in the thermal and partly diluted in the mixing region, then it is carried up in the strongly rising air from Q through P into the mixing region where it is diluted and spread out in all directions to be mixed into the entire thermal.

**1.1.1** Cumulus over the English Channel and North France at a time when the sea and land had roughly the same temperature, so that the coastline is not visible in the cloud pattern.

Newly rising towers have a sharp outline and clouds of all sizes, from tiny newly growing and finally evaporating fragments to the largest, are present simultaneously.

**1.1.2** Cumulus produced by a grass fire in Uganda. The smoke reaches up to the base of the stable layer, and only where cloud is condensed does the smoke penetrate further. The cloud is made more likely by the water vapour formed by the combustion and by the additional heat from the fire which enables the thermal to rise further in the stable layer. There were no ordinary cumulus in the neighbourhood.

**1.1.3** Cumulus formed in the plume from a forest fire seen from Boulder, Colorado. In this case the cumulus towers above the fire penetrated far enough for a column of white smoke to be left well above the rest of the smoke when the cloud evaporated. In the foreground are some small cumulus formed naturally over the mountains.

The wind has little shear and carries the smoke eastwards from the fire.

**1.1.4** Close up view of a cumulus top showing the mixing region of a rising thermal, seen from an aircraft about 300 metres higher. The oblique sunshine casts shadows on the lumpy details.

**1.1.5** Cumulus growing inland over the southern tip of Arabia. A sea breeze is blowing and the air only becomes warm enough for the thermals to rise to the condensation level about 3 miles inland.

Although this area is desert, dry channels left by occasional heavy rains mark the whole land surface. Most of the year the upward growth is stopped by a stable layer and no rain falls.

**1.1.6** Cumulus in central Mexico showing a rather high cloud base (2,600 metres) in a relatively dry air mass. The wide bases, some with rather smooth outlines, suggest that the clouds are probably located in the crests of waves in the airflow over the mountains. The cloud is almost castellatus for much of it consists of towers sprouting from these wave clouds (see 1.4).

**1.1.7** Cumulus over the Bardai Oasis (Tibesti, Sahara) at about midday. The upper air is both dry and stable. The clouds are flattened when the thermals reach a sufficiently stable layer (see 1.6) but are nevertheless rapidly evaporated as they spread out. The mixing causes loss of buoyancy which itself induces further mixing motions. Often the air is too dry for any cumulus to be formed (the condensation level is then too far above a stable layer).

**1.1.8** Cumulus which grow rapidly have sharply outlined tops in which condensation is occurring on all the nuclei available. The bases of these cumulus are at a uniform height which is the condensation level of the well-mixed layer below. When the ground is wet, or early in the morning, or over very sloping ground, the base is often very variable in height. This scene is over the plain at Boulder, Colorado.

1.1.1

1.1.2

1.1.3

1.1.4

1.1.5

1.1.6

1.1.7

1.1.8

**1.1.9** This is a view of the southern half of Florida looking southwards from a satellite at a height of about 160 km. The East coast is seen in the lower left corner and the West coast opposite.

In the centre cloud is absent over Lake Okeechobee (see 1.1.15) which is not significantly warmed by the sunshine. The cumulus are arranged in streets (see 1.3) over part of the peninsular and the wind is from the west, blowing along the streets. Many of the clouds can be seen to be leaning over in the same direction as the wind. The streets are not developed in the lee of the lake, and the clouds begin to form some distance inland in the west, but continue out to sea in the east.

In the distance are sheets of anvil cloud (see 1.6.4).

**1.1.10** The water close to the east coast of Florida near Jacksonville is cooler than the Gulf Stream which flows northwards some distance out to sea. The wind is almost parallel to the edge of the warm water over which cumulus is formed. This is a satellite photograph from the same flight as (1.1.9), and is taken looking northwards.

**1.1.11** That cumulus grow preferentially over land in sunshine is seen well in this stereo view of small islands in the Aegean sea. The cloud towers in the centre of the picture are leaning over towards the east. To the south of the island is an area of small cumulus which are small castellatus (see 1.4); they show how much smaller the clouds are when they originate in small regions of condensation and do not arrive at the condensation level as thermals from below.

**1.1.12** Cumulus growing over the Canadian Rockies in early summer at about midday. Some of the towers are very small and very quickly evaporate; others, which probably originate from ground with less snow cover, and are warmer, produce taller towers. All are rapidly evaporating and leaning over towards the ENE. The tower in the centre is growing rapidly and has a sharply outlined top.

**1.1.13** Cumulus towers closely packed where the cloud amount is growing rapidly over land in the morning (Connecticut). Even so, in 3D it is possible to see that between the towers there are still very large spaces. In the extreme distance on the right some towers are becoming glaciated. The growing towers are sharply outlined.

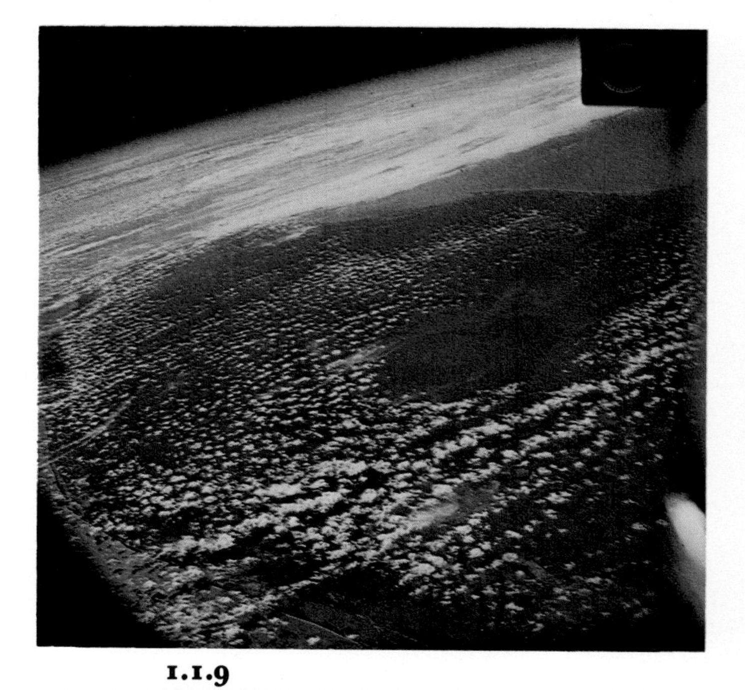

1.1.9

1.1.10

1.1.11

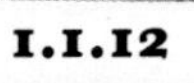

1.1.12

1.1.13

**1.1.14** This shows the east coast of Florida with cumulus growing inland with the wind from the east and increasing with height so that the clouds lean towards the west.

In the distance cumulus can be seen over Cuba with cirrus above. (Photo from about 9,000 metres.)

**1.1.15** Cumulus absent over Lake Obeechobee (see 1.1.9) about midday seen from about 9,000 metres. There is rather little wind and so little tendency to form streets of cumulus. This picture was taken just before 1.1.14.

## 1.2 Pileus

Pileus are smooth cap clouds formed temporarily in the air lifted above a rising cumulus. They are occasionally seen above thermals which are still below their condensation level, and they then look like small wave clouds.

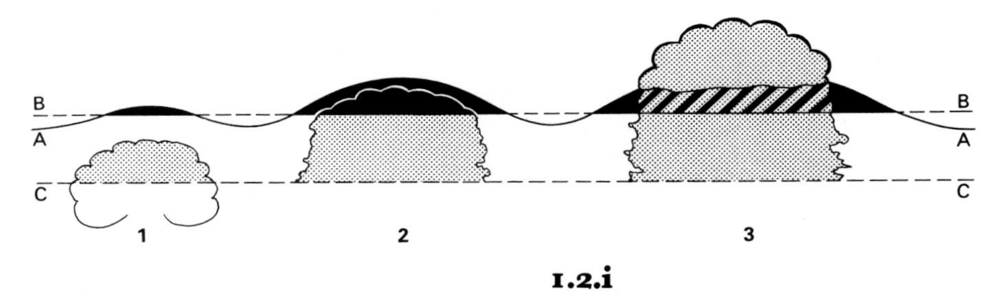


**1.2.i**

**1.2.i** illustrates three stages in pileus growth. Originally there is a stable layer at A and the air just below has a condensation level at B. The thermal has a condensation level at C, and as it approaches A some of the air is lifted above B. The pileus cloud is indicated in black. Occasionally C coincides with B so that the thermal is not visible as cloud.

In stage 2 the cumulus appears to have a very smooth top. Sometimes, but not always the thermal is buoyant enough to penetrate the stable layer, in which case the pileus cloud enters the mixing region and the cumulus appears through it, often leaving a skirt of the original pileus surrounding it for a minute or two. Any unmixed pileus cloud must ultimately fall back to level A.

**1.2.1 (a)** A wide view of a large cumulus growing over Umtali (E. Africa) with pileus just reached by the cumulus top

**(b)** the same a few minutes later with the cumulus just beginning to penetrate the pileus.

**(c)** the rising part of the cloud has now widened, and with it the pileus.

**(d)** the cumulus is halted at the stable layer over a wide area. In the distance are glaciated anvils (see 1.6 and 2.2) and it is possible that the pileus has also become glaciated and therefore persists like the anvils.

1.1.14

1.1.15

1.2.1a

1.2.1c

1.2.1b

1.2.1d

**1.2.2** Pileus, near London, in strong wind shear displaced downwind from the thermals producing them. On this occasion some pileus formed before the cumulus was visible below it.

**1.2.3** Pileus covering the whole top of small cumulus which do not penetrate it. Some waves are also forming in the layer of the pileus due to motion over hills.

## 1.3 Streets

Cloud streets are lines of cumulus cloud along the direction of the wind. Strictly speaking, when there are several equally spaced streets they lie more nearly along the direction of the wind shear: the air motion consists of thermal upcurrents concentrated under the cloud lines with weaker sinking motion in between.

In order that cloud streets shall occur in the first place it is necessary that the convection be limited to more or less the same depth of air over a large area. The streets are spaced at two or three times the layer depth. Thus, in the early morning thermals arise with a chaotic distribution over a fairly flat land area; but when the convection reaches up to a stable layer, if there is wind shear, the thermals become arranged in streets. This is so even if clouds are not formed. Later in the day, if the convection penetrates the stable layer the arrangement of clouds becomes disorderly again, perhaps being concentrated over specially good thermal sources such as hills or where showers are occurring. For cloud streets to persist it is important that they should be rather rapidly evaporating, otherwise the cloud spreads out at the stable layer to form a complete sheet of stratocumulus (1.6); consequently streets are more common and long lasting in a subsiding air mass. If the depth of the convection layer is changing streets do not become established with a uniform regular spacing.

Single streets are often seen downwind of strong thermal sources or along the edges of heated areas when the wind blows along that edge (1.1.9).

**1.3.1** is a single street of cumulus aligned over mid ocean north of the Chagos Islands, probably along the edge of an air mass cooled at the surface by downdrafts in showers (see 1.5.4). The individual clouds are growing rapidly and will soon rain; then the street will break up because the downdrafts tend to align the cumulus more nearly across the wind shear (see 3.4).

**1.3.2** shows parallel streets over land in which the spacing is determined mainly by the height of the cloud base. There is no uniform upper limit reached by the cumulus and the streets are not very well formed and are raggedly spaced. The clouds are evaporating rapidly, but as soon as a few cumulus grow larger the streets will disintegrate.

**1.3.3** illustrates a shortlived phase in the dispersion of a layer of stratus when the sunshine became suddenly stronger on the clearance of upper layers of cloud. The lines lay along the wind (and wind shear) while the convection was limited uniformly by the stable layer at the top of the original layer of cloud.

**1.3.4** is a picture of fog streets. The shallow fog is being rapidly dispersed by morning sunshine, and so long as the convection is uniformly limited by a stable layer the thermals become arranged in streets by the wind shear.

**1.3.5** This stereo picture was taken looking northwards over central England (Oxfordshire) in a N.E. wind in a summer anticyclone in mid morning. The haze top marks the stable layer which is only a small height above the condensation level. The shallow clouds are rapidly evaporating and because the convection had the same upper limit over ground of fairly uniform height the streets became uniform over a wide area.

It is uncertain how the cloud shadows influence the convection: in a case like this with oblique sunshine the effect could be strongly to perpetuate the street arrangement.

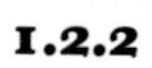

1.2.2

1.2.3

1.3.1

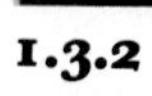

1.3.2

1.3.3

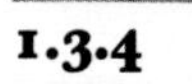

1.3.4

1.3.5

**1.3.6** Streets are common over the open ocean; these are typical of vast areas of the tropical oceans where cumulus grows almost throughout the year. Sometimes the sky is packed with cloud, at others there are only a few rather isolated streets. When streets occur there is usually no rain, or almost none, and the convection rather feeble, particularly below cloud base. The usual requirements of some wind shear and a layer of uniform depth apply here; when there are so few clouds, but streets nevertheless, the street pattern probably exists in the air below where there are no clouds. Birds often soar in such streets.

**1.3.7** was taken from a satellite looking southwards on the same occasion as 1.1.9 and 1.6.7 (which it overlaps). In the centre is the western end of Bahama. The shallow water of the Bahama bank is paler than the deep water of the Florida Strait. The largest streets extend a small distance downwind (westwards) from Bahama Island.

The cloud shadows are best seen on the sea when there is most intense specular reflection of sunlight.

**1.3.8** The clouds in these streets have a depth which is small compared with their height above the ground and are clearly limited by a stable layer. They lie E-W over Florida about two and a half hours after sunrise in August and the streets extend over the Gulf coast. Far to the west the convection is producing rapidly growing cumulus. In the right centre of the picture lines across the main street direction are clearly seen, the spacing being less. Possibly this is due to a change in the direction of the wind shear at cloud base, for all the clouds have a tendency to lean southwards in this area. The narrower spacing is related to the shallower depth of the cloud layer. The view is from about 10,000 metres.

**1.3.8**

**1.3.9** These streets over Florida, taken on another August day are composed of much more rapidly growing clouds, and where the growth is least uniform the lines are least well developed. In the foreground the clouds are appearing for the first time along the street lines, showing that the street motion had developed in the air below the condensation level.

**1.3.9**

1.3.6

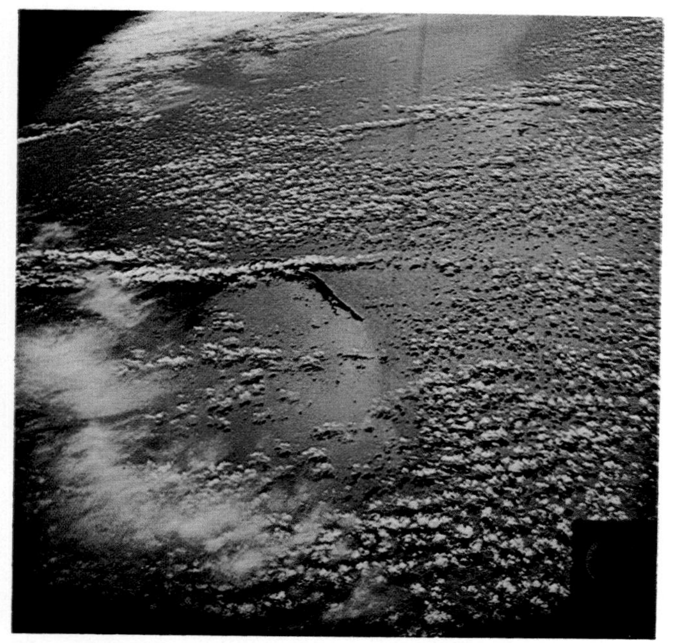

1.3.7

## 1.4 Castellatus

If the air is unstable for saturated air but stable when dry, thermals often sprout rapidly upwards from thin clouds formed in waves (5.5) or for other causes. The cumulus is then small compared with its height above ground (see 1.1.16), and is called castellatus because of its turret-like shape. More strictly castellatus is defined as cumulus which does not arise from thermals ascending through the condensation level, but which derives its buoyancy solely from condensation.

**1.4.1** is very typical. The castellatus grows from bases that are wider and smoother—almost wavelike.

**1.4.2** shows floccus, in the upper right quadrant, which is castellatus which evaporates very rapidly after swift upward growth as narrow towers. In the distance is seen a line of castellatus growing probably along the edge of a moist patch of air which has been lifted only just through its condensation level.

**1.4.3** is less obviously castellatus. Among these mountains in Rhodesia is stratus being dispersed and carried upwards in cumulus in the distance, by the sunshine on the mountains. At the highest level are smooth wave clouds (see 5.1) and the wind shear evident in the middle level clouds makes it probable that much of it is initiated by wave motions; but it is clearly reinforced by the cumulus rising from below, of which some evaporates on mixing into the drier air above the stratus and then recondenses at the higher condensation level.

The general appearance of the middle level indicates rapidly sprouting cumulus tops which are castellatus growing on cumulus.

1.4.1

1.4.2

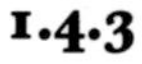

1.4.3

**1.4.4** In the evening when the cumulus which grows over mountains by day begins to evaporate, a sinking motion is produced which temporarily produces ascent in air over the nearby low ground. This example, at Fort Collins, Colorado, close to the Rocky Mountains, half an hour before sunset shows the bright white sharply outlined tops of the growing towers, illuminated through clear air above the condensation level. The bases are dark because in the low sunshine shadows are cast on the lower parts of the clouds.

In the distance on the left is a shower cloud still persisting over the nearby mountains.

**1.4.5** Castellatus is common, though less prominent, in deep lows in which horizontal convergence is strong. The air is carried by the convergence up to the condensation level where, under overcast skies and with no heating from below, towers of castellatus sprout. This view was in the early morning on the west coast of Denmark. The cloud is typical of cyclones in which the heat is derived from widespread cloud growth, rather than mainly in concentrated frontal cloud. In tropical cyclones the wind shear distorts the cloud arrangements (see 14.4.4-5).

**1.4.6** The rapid ascent of the towers produces a rather exciting distortion of the much wider base clouds of castellatus. Often dramatic lighting contrasts and twisted cloud fragments adorn the quickly changing shapes.

**1.4.7** When the ascent of thermals is very rapid the behaviour is more like a thermal in neutral surroundings which leaves no tower behind it. The castellatus, in the middle left of this picture particularly, are showing this effect. This is a form of floccus, but the rapid disappearance of the tower is because there never was much of it.

**1.4.8** At higher levels in a cold low, which is a cyclone maintained by the continual generation of heat by clouds, castellatus is widespread. In stereo the narrow towers, are seen rising from tenuous bases or penetrating stable layers at which residue is deposited by less strong thermals. Note that the uppermost towers do not correspond well with the towers below the moist layer and that there are turrets below in the foreground. This was about an hour after sunrise over Stafford, England, in July.

**1.4.9** The castellatus in the middle of this picture show that there is quite a large shear at their level. In other parts of the sky the wavelike structure of the cloud patches is apparent. Most castellatus is formed in rather small amplitude waves, and so very little shear is generated in them (see 5.8 and 6.1), but evidently here there was already shear above the wave cloud level. This picture was taken with a very wide angle lens so that the top of the picture is at an elevation of about 75° above the horizon.

**1.4.10** Cumulus castellanus sprouting in tall towers in the Caribbean. The buoyancy originates almost entirely in the condensation, so that the updrafts are much stronger in the cloud towers than in the thermals between the sea and the cloud base.

The still rising tower is sharply outlined, but not so the evaporating fragments. In this case the surrounding air is dry enough to evaporate the clouds quickly.

1.4.4

1.4.5

1.4.6

1.4.7

1.4.8

1.4.9

1.4.10

## 1.5 Warm Rain

Rain is described as warm when no ice crystals are involved in its formation. Warm rain is very common in the tropics and over the more temperate oceans, but is rarer over land in high latitudes. If it is produced in small clouds the updrafts must be weak; otherwise cloud particles reach the upper and outer regions of the cloud before they have grown large enough (100μ at least) to avoid rapid evaporation.

**1.5.1** Warm rain over mid-Atlantic at 46°N from cumulus, some of which is almost castellatus. It is thought that the abundance of salt nuclei, when other small nuclei such as those produced by combustion are largely absent, helps in the more rapid growth of the droplets to sizes greater than 10μ after which collision begins to surpass condensation as the main mechanism for droplet growth.

**1.5.2** shows showers of the equatorial oceanic rain belt arranged in a line, presumably by the action of their downdrafts in locating new updrafts. Although there is some glaciation in the tops of these clouds the ice particles do not play a major role in the rain formation: on the contrary, the presence of larger droplets in the upper parts of the clouds is probably a major agency for the glaciation (see 2.2).

The dark lines of low cloud in the foreground on the left and in the distance on the right are probably the tops of newly growing updrafts over the downdrafts spreading out beneath the showers (see 1.5.8. p36, and 3.3).

**1.5.3** is an example of evening warm rain from castellatus over land. This cloud is one growing on the occasion of 1.4.4, about half an hour later that evening. The deep reddening below the cloud base shows that the haze was mainly confined below the cloud base (see 1.1).

**1.5.4** The thin wave cloud in the distance was a successor to 5.1.10 which was photographed about an hour earlier. During that time cumulus grew rapidly over the smaller hills and copious warm rain soon fell. When showers begin the cloud base often looks like mamma for a few moments (see 3.6.7) but soon the larger drops fall ahead of the downdraft and form rain streaks.

**1.5.5** Cumulus formed over the sea in August can have much weaker updrafts than those formed at the same time over land because of the lower sea temperature. On this occasion showers were drifting across Edinburgh from the sea in the afternoon, while inland the convection was too strong for showers.

**1.5.6** When convection becomes weaker in the evening warm rain can be formed although no showers occurred during the afternoon. On this occasion the fallout could be seen and the rainbow confirmed that there was no glaciation. The secondary bow can just be seen on the right.

**1.5.7** One of many isolated showers over the Indian Ocean within 5° of the Equator. These larger clouds are largely castellatus.

**1.6.1** Stratocumulus (see next page) occurs in large patches sometimes when the afternoon sunshine is strong enough to evaporate some of it. This is over central north France, and the unbroken layer can be seen in the distance. There was very much haze, largely man-made pollution, below the inversion which capped this layer of cloud.

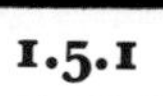
1.5.1

1.5.2

1.5.3

1.5.4

1.5.5

1.5.6

1.5.7

1.6.1

**1.5.8** Warm rain clouds disappear very rapidly after the rain has begun and a downdraft develops. The downdraft is produced partly by the weight of the rain falling into air that did not previously contain it, and partly by evaporation which cannot occur significantly within cloud because cloud droplets evaporate much more rapidly, even in a downdraft. But sometimes the rain is in clear air above cloud base, in which case that air may be brought down below cloud base by the cooling.

Downdrafts reaching the sea surface spread a carpet of cooled air over an area much larger than the original cloud, and until this has become warmed throughout by the sea no new clouds occur above it. In this picture the large holes in the cloud cover are where cold downdrafts from clouds which have now completely evaporated have spread out over the sea.

There is a tendency for new clouds to grow on the edges of a spreading downdraft, especially where two downdraft fronts are advancing towards each other.

Although the convection from the sea is required to support the clouds by carrying up the water vapour, thermals rising through the condensation level do not play an important part in individual cloud growth. Most of these clouds are castellatus.

## 1.6 Anvils and Stratocumulus

When a thermal or any other more complex upcurrent arrives at a stable layer which it has not enough buoyancy to penetrate it spreads out beneath after temporarily overshooting it. The cloud which spreads out is an anvil cloud. Once the spreading has begun, the mixing and evaporation that normally occurs on the outside of a water droplet cloud is reduced by the great stability, because the heat lost in the evaporation only produces small up and down motions; it therefore does not produce much further mixing.

Stratocumulus is the name given to a layer of cloud formed by cumulus convection from below.

**1.6.i** shows three stages in the motion; the shape of a cloud depends on where the condensation level lies in this motion pattern.

The underneath side of the spread out anvil cloud may be far above the condensation level of that air. Consequently, being an interface between dense cloud and clear air, this boundary is rather sharply defined and partakes of the movements of the air. At a condensation level air moves through the cloud boundary (see 3.6).

**1.6.1** see p34.

**1.6.2** This layer of stratocumulus seen from about 250 metres above it, was formed over the warm ocean (10°S, 13°W) with its top nearly 7°C colder than the very dry air a very short distance above it. Radiation from the top, at about 1,300 metres above the sea, causes a loss of heat with downward convection producing a lumpy appearance in the layer. Convection from the sea below is on a larger length scale and gives rise to the larger patches. The gaps are where the down motion is more concentrated.

Sometimes the upward thermals producing the cloud are strong enough to mix into them a little of the dry air above when they overshoot into the stable layer. This tends gradually to dry out the layer.

If the air is moving to warmer sea the base of the cloud rises; if to colder sea the temperature of the whole layer falls accordingly and the cloud probably thickens and the inversion at the cloud top is intensified.

Because of the large and sudden decrease in humidity above the top of the cloud, the layer below acts as a very efficient radio (radar) duct.

1.5.8

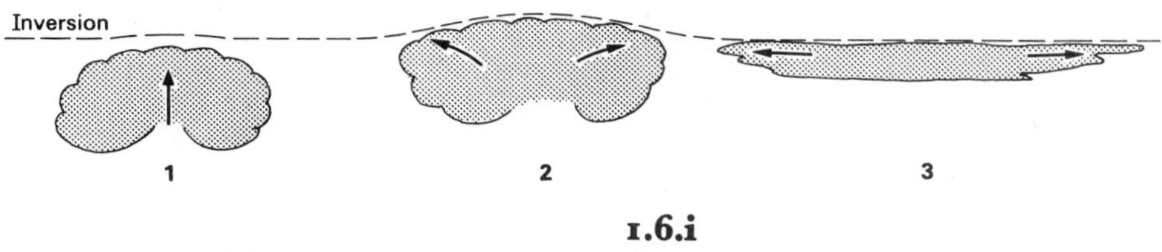


1.6.1

1.6.2

**1.6.3** Cumulus is building up over a peninsular in South Wales and is spreading out at an inversion to form anvil stratocumulus. In this case it appears that, because of the mixing in the narrow towers below, the stratocumulus is not far above its condensation level: many of the towers fail to reach the layer because of evaporation. Consequently, in the wave motion produced by the arrival of these vigorous thermals at the layer, a gap is produced where there is descent. This is particularly noticeable to the left of the towers. (See also 3.6.6.)

**1.6.4** The spreading out of cumulus at stable layers always leaves a rather patchy distribution of humidity and the humid layers are very suitable for the formation of wave clouds (see 5.1). It is often impossible to distinguish anvil stratocumulus from cloud reformed by wave motions later in the same layer. This is an early morning example near Shrewsbury after a showery night.

**1.6.5** Anvil clouds sometimes sprout towers if convergence is occurring. In a photograph of this sort it is difficult, without stereo viewing, to see that the main mass across the centre of the picture is detached from the cloud below. Part of the highest turret is glaciated.

**1.6.6** Long shadows extend from the tops of the largest cumulus across the top of the sheet of anvil cloud formed by cumulus over the mountains of southern Ethiopia. In the foreground of this picture from a Mercury satellite at 160 km. is Lake Rudolf in northern Kenya. There is no cloud over the lake. Cloud marks the eastern edge, but there is a cloud-free area about twice as wide as the lake on its western side: from here the cloud has already evaporated in the evening.

In 3-D the Earth's curvature can be clearly seen.

**1.6.7** This is a view of which the foreground was shown in 1.3.7. The edge of the Great Bahama Bank, where the shallow water is paler, is clear and the northern end of Andros Island is seen beneath the nearest patch of stratocumulus. Cuba lies beneath the main more distant sheet of cloud with Jamaica almost on the horizon near the centre of the picture. The layer is penetrated in a few places over Cuba where flattening occurs at a higher level, and although the layers are initially formed over the land, the coastline is not shown in the cloud outline because an equilibrium similar to that in 1.6.1 is established as it moves over the sea.

**1.6.8** This is a view from a Mercury satellite at about 160 km. over the west coast of Africa at 28°N. In the bay at the edge of the cloud sheet is Agadir and the south coast of Portugal is seen just below the horizon.

Convection over the warmer water forms stratocumulus in lines along the wind beneath the inversion due to subsidence as a cold air mass moves southwards. Heating over the land in the middle of the afternoon evaporates the cloud so that its south eastern edge marks the coastline. (See 9.3.)

Over the higher ground inland thermals are penetrating the inversion to form cumulus. The pattern near the edge of the cloud sheet in the west is not obviously explicable.

1.6.3

1.6.4

1.6.5

1.6.6

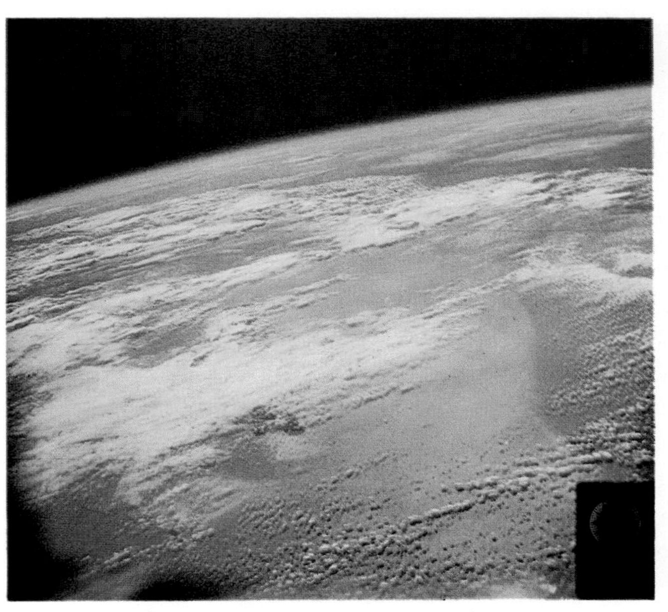

1.6.7

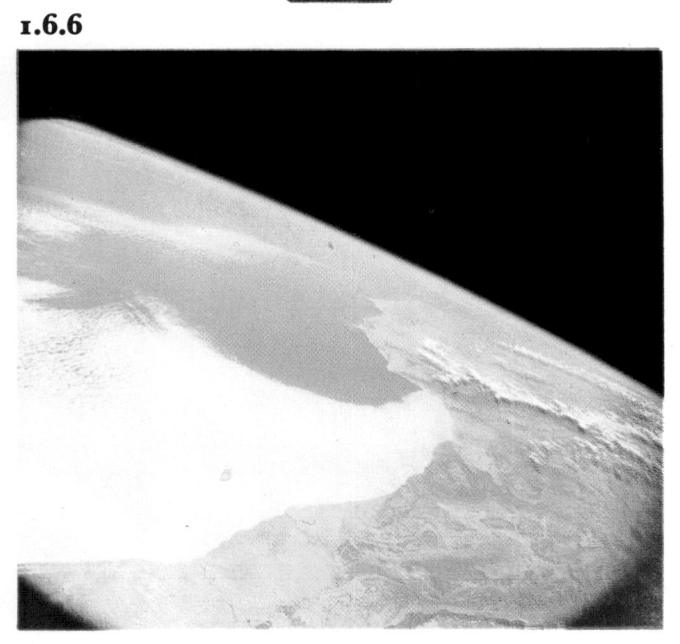

1.6.8

# 2 GLACIATION OF CONVECTION CLOUD

Glaciation is the process of freezing of cloud droplets. This does not occur until the droplets have been cooled well below 0°C; it occurs more quickly the lower the temperature and the larger the droplets.

Ordinary cloud droplets freeze almost instantaneously at temperatures below −40°C, but may remain unfrozen for many hours at temperatures above −10°C.

The effect on the cloud depends on how many of the droplets freeze and what is the humidity of the surrounding air. If only a minority become frozen they grow by condensation (sublimation) from the surrounding vapour because the vapour pressure over ice is less than over water at the same temperature, and the unfrozen droplets evaporate: in this case the frozen particles soon become large enough to fall out and the cloud is transformed into falling streaks. By this means holes appear in thin layers of supercooled water cloud.

If most of the drops are frozen, and they are also very numerous and small, as in the top of a vigorous cumulus cloud, no fallout is observed, but the cloud does not evaporate like the former water cloud. It then becomes silky, diffuse, and fibrous; it does not cast sharp shadows and may endure in tenuous patches for long periods with little apparent motion when it achieves evaporative equilibrium with the surrounding air. Because of the absence of evaporation the mixing motions characteristic of the edges of water clouds do not occur, and shear in the air can draw the cloud out into long streaks.

See Chapter 4, and sections 5.4, 7.4, 11.4, and 13.3 for other aspects of glaciation.

The difference in vapour pressure over water and ice at different temperatures can be alternatively measured as the distance below the ordinary condensation level of the ice evaporation level, which is where the air first achieves saturation for ice during ascent. Each 100 metres corresponds to a temperature difference of 1°C between the dew point and the frost point:—

| Temperature | 0°C | −9°C | −19°C | −32°C | −41°C |
|---|---|---|---|---|---|
| Height difference | — | 100m | 200m | 300m | 350m |

### 2.1 Glaciation of Cumulus and Stratocumulus

**2.1.1** This shows the diffuse appearance of the tops of glaciated cumulus over Luxemburg late on a summer morning seen from 6,000 m.

**2.1.2** This shallow autumn cumulus was formed over the hills of Kent in a wind from the N. Sea. On becoming glaciated it produced copious fallout, which is seen coloured orange because it is both illuminated and viewed through a hazy layer. The cloud itself is much whiter, being above the haze top. There are mamma (see 3.6) on the bottom of the falling snow.

**2.1.3 (a)** The large cumulus is the main contributor to the anvil stratocumulus surrounding its top. After it has begun to rain its top has become glaciated.

**(b)** There is a clear contrast between the glaciated and unglaciated parts of the anvil cloud, with fallout from the former. This is another view of the edge of the glaciated anvil of (a).

### 2.2 Glaciated Anvils

**2.2.1** The glaciated part of a cloud remains after the water droplets have all evaporated. The almost flat base of this residual ice cloud is the melting level, and the rain falling below it is not visible (sometimes it can be identified by a rainbow). The water droplets are evaporated by mixing of the cloud air with drier air around it, by evaporation in the presence of ice crystals, and by downdrafts. All of these mechanisms can be seen in action on occasions like this one.

**2.2.2** Large cumulus often produce showers of copious rain over the desert areas of low latitudes, but in Tibesti (Sahara) much of the rain sinks underground and the rest evaporates within a day or so. The storms shown here produced lightning throughout the night. Probably convection from the higher mountains had penetrated the stable layers which prevent the development of cumulus below 3,000–4,000 metres over much of the Sahara.

**2.2.3** A glaciated anvil spread out by wind shear and remaining unevaporated is common on the top of a shower cloud. The hook shape on the left shows that the spreading is fastest in the middle of this arm of the anvil.

On the edge of the anvil, in the right picture in particular, are fragments that often look like water cloud indicating, probably, new condensation. Such fragments disappear rapidly by evaporation unless quickly glaciated.

There are mamma falling from the thicker part of this anvil, and because of the copious fallout the cloud base is not easily identified.

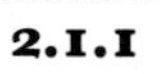

2.1.1

2.1.2

2.1.3a

2.1.3b

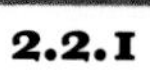

2.2.1

2.2.2

2.2.3

**2.2.4** The tallest and largest of the cumulus towers penetrating the layer of anvil stratocumulus has become glaciated on the left, and its appearance is transformed. (Central Sweden).

**2.2.5** This large storm cloud has many typical features well displayed. The new growth was mainly on the side towards the photographer, and a series of anvils were formed and drifted away, spreading out, glaciated, while new towers rose in succession, each penetrating the near edge of the anvil of its predecessor. In the state shown, the nearest anvil is spreading at about its maximum rate, and it is temporarily producing some protuberances like those in 2.2.3, with the appearance of new water droplet cloud. This cloud produced a tornado beneath its later rapidly growing towers and was remarkable for being the only violent cumulus growth over a large area that day (July, South Dakota).

**2.2.6** These anvils were observed near the Shetland Isles looking from 14,000 metres towards Norway, over which the most distant anvils on the right were situated. New thermals arriving at the tropopause protrude inside each ring of glaciated anvil cloud.

## 2.3 Anvils and Plumes

Glaciated tops endure at high altitude for hours and usually indicate rain. In Central Australia the cattle on the larger stations seem to follow storms visible at a distance and feed on the vegetation that flourishes during the next few days.

Some anvils become completely glaciated, in which case the particles remain small and do not appreciably fall out. They may therefore be carried for large distances downwind of any place where large cumulus grow preferentially.

Perhaps the most notable anvil plumes are those from oceanic islands because they can be seen from very large distances. Undoubtedly over the centuries boats in the Pacific Ocean have been navigated by following these tall clouds and their evening trails. The plume growth has recently been observed from satellites also.

**2.3.i** In this diagram it is implied that the cloud is growing over higher ground. There are other sources, such as oceanic islands or even large industrial areas which serve to locate the cloud growth in one place.

If the wind increases markedly at the tops of the cumulus where the anvil spreads it will be carried off as a plume extending several tens or even hundreds of miles downwind if it has been glaciated. But even if the wind increases gradually the appearance may be very similar when the updrafts in the cloud are strong.

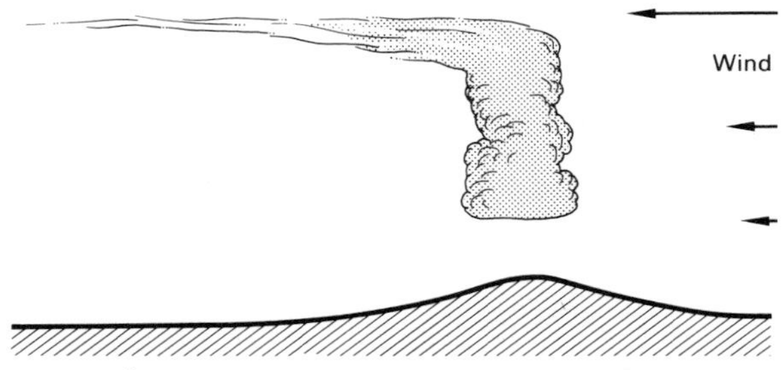


**2.3.i**

**2.3.1** The various stages in the growth of a large cumulus are seen together over the Santa Maria mountains in Western Arizona. The oldest tower has a glaciated anvil which is sheared over to a length of about 20 km., while the tower on the right which has newly risen to these levels is only just beginning to show the presence of the shear.

It is probable that cumulus is not growing over the lower ground because there is a stable layer below the condensation level which only the warmer thermals from the higher ground can penetrate. If the air is stably stratified at levels below the higher ground there will also be an anabatic flow of the heated air towards the higher ground. (See 9.2).

There are mamma on the overhanging anvil, in this case composed of falling ice particles. At the top of the anvil are fragments of cloud which were completely glaciated rather quickly so that the smaller droplets also froze: these persist without any perceptible falling of the particles through the air, as tenuous filaments of cirrus drawn out by the wind shear.

At the centre of the cloud mass between the two largest towers is a smaller glaciated tower which has lost its sharp edge: the tower on the right is still rising and is very sharply outlined. (See also 3.4.2.)

2.2.4

2.2.5

2.2.6

2.3.1

**2.3.2** The Mercury satellite which also obtained 1.6.6 took this picture over the southern Sudan. The nearest anvil is very similar to 2.2.6, while the more distant ones have anvils which completely cover a large area. A few new towers protrude on the right with shadows cast by the low western sun on the anvil cloud. This, like 1.6.6, 14.2.1, 14.3.1 and 14.4.1-3 are among the first stereo pictures recovered from a satellite in orbit.

**2.3.3** In this example the plume of anvil cloud becomes more predominantly cirrus in form as it is carried further from the parent cloud. It was photographed at about midday in September near Gabes (Tunisia) looking eastwards towards Jerba. The parent cloud was almost stationary.

### 2.4 Ice Fallout

**2.4.1** The melting level is visible when the sun is behind the falling snow. The rain, which the snow becomes on reaching the melting level, obscures light much less.

**2.4.2** Isolated thermals form cumulus towers. The residual water cloud left by this one has evaporated leaving only the falling frozen particles. The sun is behind the cloud which has a sharply outlined top because it is still rising.

# 3 SHOWERS

When rain occurs and the cloud tops are glaciated there are several new kinds of motion to be observed which are either less well developed or non-existent in cumulus which are not frozen or raining. These range from small waves and mamma, through motions on the scale of the whole cloud including those which regenerate it, to much larger ones which organise showers in cold fronts and cold lows.

## 3.1 Waves and downdrafts

Wave motions are produced in the neighbourhood of shower clouds by the strong up and down currents in them. We have remarked upon those in a very stable layer when an ascending thermal reaches it (1.2 pileus, 1.6.3 holes in anvil stratocumulus) and the effect of rain downdrafts in subduing convection (1.5.7). Around stronger storms many other wave motions are evident.

**3.1.1** The arrival of new towers at a stable layer produces wave motions which are sometimes revealed as thin wave clouds which last only for a minute or two. Below such clouds in this picture we see the edge of the main glaciated anvil with small patches of water cloud similar to those seen in 2.2.3 and 2.2.5 on the anvil edge.

**3.1.2** There is often a stable layer just below the condensation level, particularly over the ocean (see fig. 1.1.iv), and when there are scattered showers about the wave motions are often revealed by small pileus-like clouds at that level. These are in mid-Atlantic, far from any mountains, yet they have some features of mountain wave clouds.

Waves of this kind from storms can sometimes be detected hundreds of miles away by barographs, though not usually by clouds. In those cases the rain is usually generated in castellatus-type clouds which rain down into a layer of air below, which is beneath a stable layer and is moving with a significantly different velocity. The downdraft air acts as a blockage to the lower layer and the impulse is transmitted over large distances as a gradually dispersing wave train, mainly in the direction opposite to the direction of the shear between the two layers. This is because the wave is very efficiently trapped when propagated in that direction (the energy is not propagated upwards and so the wave is much more slowly attenuated).

In this picture the clouds are about two miles from the nearest rain: the sky is almost completely overcast with spread anvils.

2.3.2

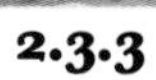
2.3.3

2.4.1

2.4.2

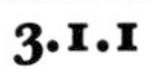
3.1.1

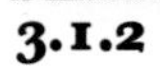
3.1.2

**3.1.3** When the rain is produced in castellatus-type cumulus, whose location is not determined by thermals originating at the ground, there is no mechanism whereby the shower clouds can organise one another, and each produces a small shower. This shows the rain streaks of about seven separate raining cumulus in the Colorado Grand Canyon, and is rather like a collection of the clouds of 7.4.1 or 2 enlarged.

When rain falls through several thousand metres of clear air a strong downdraft is produced by the weight of the rain and by the cooling due to the evaporation of some of it. The air cannot be cooled below its wet bulb temperature by this means, but even so the cooling may exceed in magnitude the excess temperature in the warmest of rising thermals. Downdrafts of 20 metres per second have often been observed beneath large showers by glider pilots. Sometimes the evaporation causes the disappearance of the rain before it reaches the ground.

When downdrafts impinge on mountain sides a rather shallow torrent of cold air down the slope is often produced. The arrival of the front of the cold air can be very sudden, producing squally winds from a direction quite unrelated to that which blew before it. In mountainous country these squalls can be observed perhaps up to 50 km. from the rain storm. They often blow up dust in dry territory, and damage windows and doors left open in the balmy air which is suddenly replaced by the cool gust.

## 3.2 Scud

Scud is fragmentary cloud below the main cloud base which is produced in rising air with more than average humidity. The extra moisture is obtained in two possible ways:—

1) by the evaporation of rain
2) by the evaporation of water on the ground.

The second mechanism is probably more important in most cases. For example, after a cold front has passed the wet ground makes the air near the surface much moister than that above. Thermals rising from it condense cloud (scud) below the main condensation level; but often such scud evaporates before reaching the main cloud base because of mixing with the intervening dry air.

**3.2.1** The scud seen beneath this storm cloud is fairly typical. It looks bright or dark according to the illumination of it and of the background. If it is produced by the lifting of air, previously moistened by rain, at the advancing front of the cold air of another downdraft, it can condense very close to the ground. Often tornadoes are started in this way (see 14.3.4).

The nearest fragment has an arched top as if it were being lifted from below, like pileus, by either a thermal or a downdraft front.

**3.2.2** A passing cold front leaves the ground wet, and often relatively warm. Consequently, even under an almost overcast sky the lowest layers become very much more moist, and scud is formed below the main cloud base and mountains are shrouded in it.

## 3.3 Squalls

Squalls are essentially produced at the advancing front of downdraft air spreading out on the ground. We have referred (in 3.1.3) to the squalls at the front of cold air accelerated down a slope. Here we are mainly concerned with squalls in air brought down to the ground by rain-cooling from a level where the wind is different. The front of the downdraft spreads along the ground, in the direction of the wind shear relative to the air at the ground.

**3.3.1** A shower is moving into England from mid-Wales in the afternoon. The rear part of the anvil with its outline blurred by fallout is seen, and beyond it a newly rising tower, on the left.

This and subsequent pictures in 3.3 must be interpreted having in mind that 3.3.i is a simplified 2D representation of systems that are always more complicated and 3 dimensional.

3.1.3

3.2.1

3.3.1

3.2.2

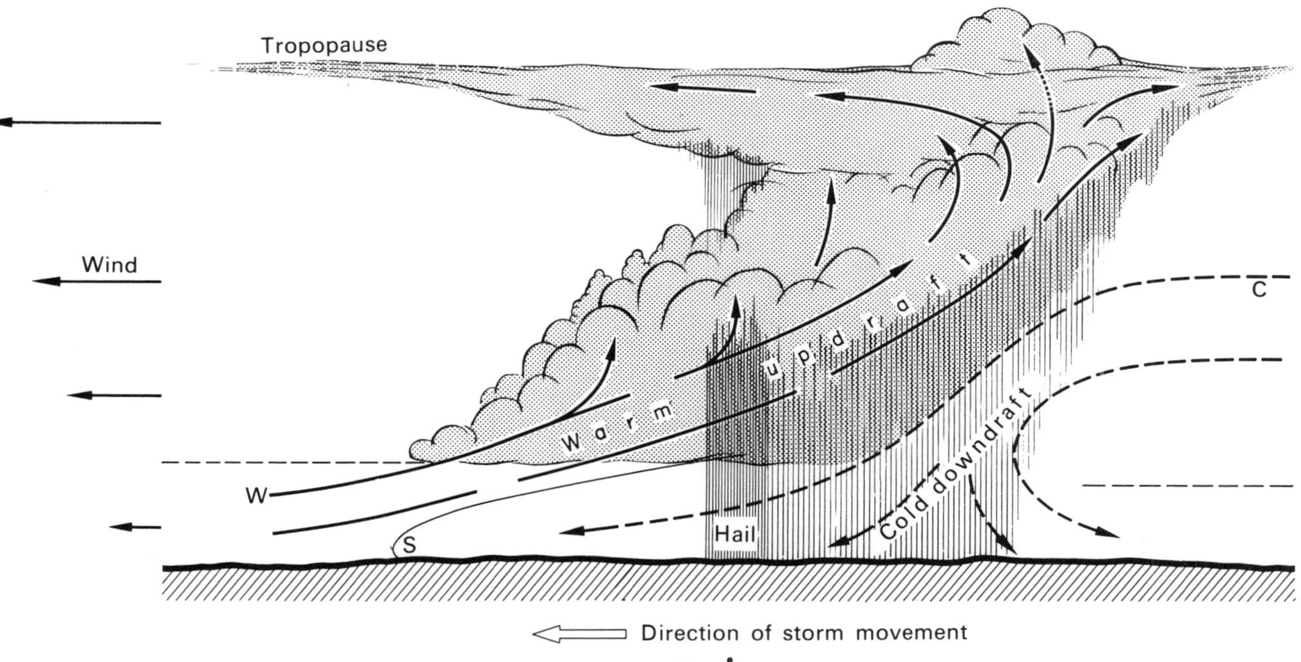


**3.3.i**

**3.3.i** is a cross section of a self-propagating shower cloud in its simplest form. Shower clouds do not usually have this simple two-dimensional structure, and the stereo picture 3.3.5 shows a twist due to changes of wind direction as well as wind strength.

The rising air, starting at W, is lifted by the advancing cold squall S, the cold air being below the continuous line. The new cumulus growth on the left develops gradually into the main updraft which is inclined so that the rain from it falls out above the main cloud base into the air at C which is overtaking it. This air is cooled and caused to descend, spreading out on the ground and leaving a carpet of cold air beneath an inversion (shown by the thin dashed line on the right) and spreading forward to regenerate the squall. (Continued on p. 48)

The effect of the shower is to exchange warm moist air at the ground (W), which is often beneath an inversion through which thermals do not penetrate until the squall lifts it, with dry air (C) which becomes cold in the downdraft and therefore appears at the ground as cold air which is also rather moist. More notable than the upward transport of heat is the upward transport of water. The updraft spreads out as an anvil, occasionally overshooting the tropopause, and the glaciated part is carried predominantly forward by the strong upper wind.

From the anvil there is often a copious fallout of ice particles. Some of the larger ones fall back into the downdraft and become the nuclei of hailstones which can only grow on particles large enough not to be carried rapidly into the completely frozen part of the cloud at the top. Hailstones require a cloud of supercooled water droplets which they can capture as they fall through it, and the only part near the top of the cloud not completely frozen is where a strong thermal has risen rapidly from below.

From the rest of the anvil the smaller falling ice particles produce mamma (3.6). Mamma may be observed on the fallout on the rear (right) side of the anvil also.

**3.3.2(a)** This view, looking eastwards, shows a hailstorm in September over the Great Plains of Eastern Wyoming where the wind near the ground was moving slowly from the south while the upper winds were strongly from the WNW. The fallout at the back (west) of the storm is illuminated by the sunshine and becomes less clear below the 0°C level where it is all rain.

The glaciated part of the anvil is carried away predominantly in the direction of the upper wind at around 11,000 metres and some updrafts are producing domes of cumulus above this. There are other anvils visible to the north.

**3.3.2(b)** is the same storm seen more closely, looking NE-wards. The dense layer at the anvil level may be partly formed like pileus. Most notable are the mamma on the overhanging cloud at about 6,000 metres above the main region of fallout. The ground is at about 1,600 metres above sea level.

The ground is darker where it is still very wet from the rain, and this darkening is easily distinguished from the cloud shadows which are sharply outlined. Just below the middle of the picture to the left of the cloud the white spots are where hail still remains unmelted.

The wind structure is basically that of fig. 3.3.i.

**3.3.3** Radar pictures of an advancing squall. The left picture shows a storm as seen in plan at Dunstable by radar at O. The circles are at intervals of 5 miles and N is in the north. The echoes near to O are produced by buildings, trees etc.

At a distance of about 8 miles in the direction OA is a bright arc which marks the cold nose. The echo is mainly from birds (chiefly Swifts) soaring in the updraft of warm air which contains an abundance of insects, and it is seen as two clearly defined arcs originating from separate downdraft centres. At about 13 miles along OA is seen a strong echo from raining cloud. This is probably the area where the updraft is developing most rapidly: often the cloud growth is in a series of spurts and the subsequent rain produces the next spurt. From about 19 miles outwards there is more rain.

The right picture is a vertical section along the direction OA taken 9 minutes earlier, when the squall was at about 11 miles distant. The cold nose is advancing at about 20 m.p.h. The vertical lines are at 5 mile intervals and the dark horizontal ones at 5,000 ft. intervals. The bright line at the bottom marks the tangent plane to the earth at the radar station, and this is well above the ground at 50 miles.

The echo is highest, reaching about 32,000 ft. at about 17 miles from O, and this is where the new towers are reaching well above the main level of spread of the anvil. The anvil spreads to a distance of 50 miles in this direction, and this line of storms could be described as a cold front.

Echoes similar to that on the squall front seen here are often observed on sea breeze fronts, where birds find insects plentiful.

3·3·2 a

3·3·2 b

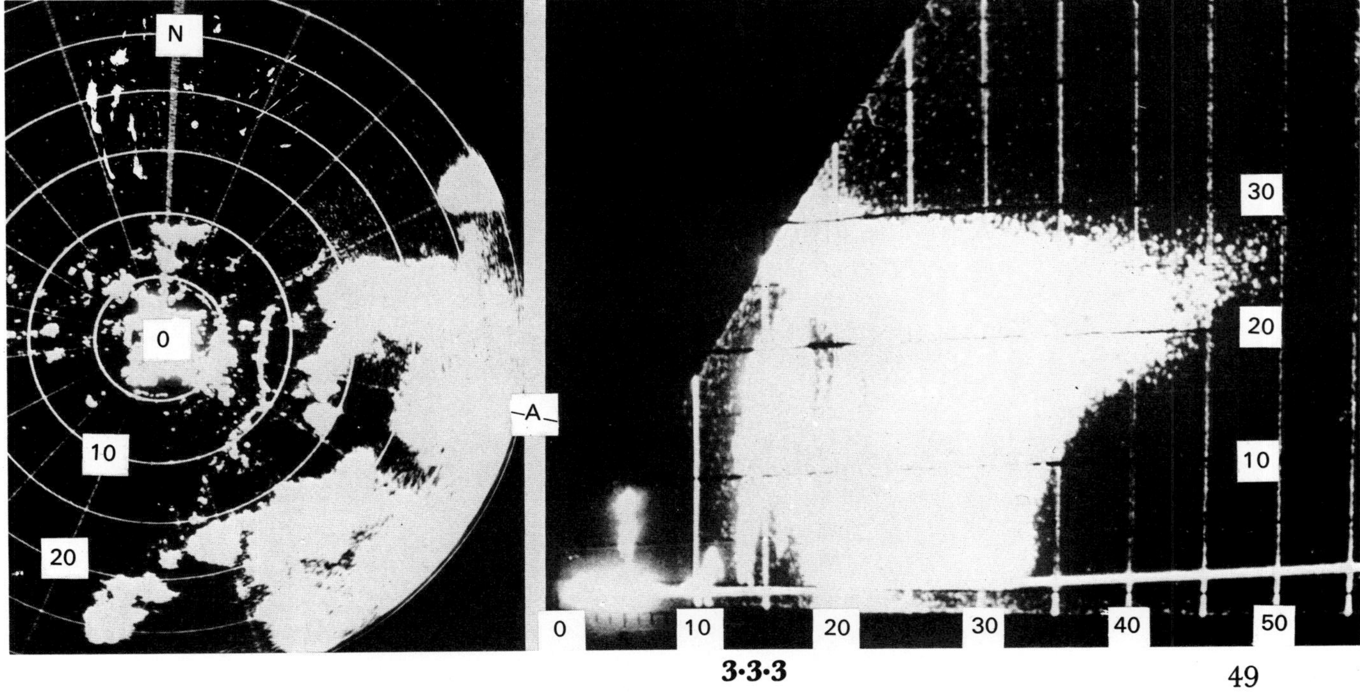


3·3·3

**3.3.4** shows the front edge of the cloud lifted above the squall advancing out of the main storm cloud on the right. It is passing an Air Force base in Florida.

Close examination shows a second layer of cloud above the lowest, whose smoothness indicates a very rapid rate of ascent: the convection in the cloud has not produced velocities strong enough to distort the cloud in the time taken for the strong updraft to produce it. (See, by contrast, 1.5.2 and 3.6.1).

**3.3.5** The rain in a self-propagating storm is here seen advancing across the plains of Minnesota. The downdraft arrives ahead of the rain and some scud can be seen rising into the cloud base in the lifted air. The cloud base is lowered a little also where it is raining. The texture of the main cloud is smooth because of the rain and ice particles falling from the anvil higher up and obscuring the new growth from view.

It is probable that earlier showers had produced the variable humidity which is responsible for the scud.

**3.3.6** The main updraft into this anvil can be seen as a powerful column of white cumulus on the left, rising out of the cloud growing up from the bottom left of the picture. The anvil overhangs towards us and from it copious snow is falling, producing the downdraft of air moving towards the left, which on spreading forward on the ground lifts the lower air into the growing cumulus that feed the column.

Above the anvil streaks of cirrus form delicate pileus. In the stereo view the manner in which the glaciated edge protrudes beyond the cumuliform top is clear. There is fallout with some mamma from this glaciated part, which gives the underside a gentle curve because of the spreading motion. The shadow of the anvil on the column is clearly defined. Although 'solid' looking in a single picture the delicate cloud is seen to be sleek and twisted in 3D.

**3.3.7** The squall line of a cold front is often best developed over the uniform surface of the sea. This one has just passed over the ship.

**3.3.8** A line of showers constituting a cold front moving away to the left early in the morning in Hampshire. Note the anvil mamma.

## 3.4 Cold Fronts

Cold fronts are of many kinds. In extratropical cyclones they are clearly marked boundaries of advancing polar air masses, but the mechanisms of the showers in them may be reinforced by those illustrated in figure 3.3.i. On the other hand, these mechanisms may cause the alignment of showers along a front across the wind shear when the downdrafts produced by different cells overlap (as in the left picture of 3.3.2), and a line of heavy showers may develop many of the characteristics of an air mass front after a few hours. This is illustrated in the 'squall line' 3.3.7 and the 'front' 3.3.8.

In the tropics the cold fronts, carried equatorwards as cold air masses enter the trade winds, are only present in the lowest very few thousand metres because the air mass has become very shallow as a result of subsidence. Such fronts are sometimes the initiators of lines of showers when they reach a region, on the west side of an ocean, where the clouds are growing vigorously up through the subsidence inversion.

The intertropical convergence is a front in the sense that, being a line of convergence it is the demarcation between air masses of northerly and southerly origin. If the downdrafts have the effect of initiating new updrafts the front tends to behave like a line of squalls, but its more usual behaviour is to die out and be regenerated at a distance of 50 to 100 km. at intervals of a day or two.

**3.4.1** is a dry cold front. The advance of the polar air was accompanied by no rain, merely the clearance of the rather drizzly cloud of the warm sector, and the most notable feature here is the clearance of the uppermost cloud. In this view, looking westwards at Reading, the cirrus was moving strongly from the west along the front and slowly southwards with the movement of the front. Because of the slow movement the cumulus which grew over the sunwarmed ground was not far behind the clearance of the cloud.

At the surface the arrival of this cold front was accompanied in mid morning by a considerable rise in temperature, but this would not have occurred under the same conditions at night. Showers occurred that afternoon about 150–200 km. behind the front; close to the front the subsidence of the cold air had made it too dry.

**3.4.2** shows the clearance looking NW at Ostende behind a cold front whose passage produced about 90 minutes of moderate rain. This front too was slow moving and the main wind component at high (jet stream) levels is along the front. The anvils of the cumulus growing in the cold air which reach up to this level are carried rapidly NE-wards.

3·3·4

3·3·5

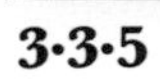

3·3·6

3·3·7

3·3·8

3·4·1

3·4·2

**3·4·3**

**3·4·3** This is the Intertropical Convergence seen looking SW-wards from 8,000 metres towards the N coast of Guyana in South America. The wind is from the East but decreases with height so that the clouds lean over towards the east. The larger clouds in the extreme distance are over the land which is about 100 km. away. Most of the rain is warm rain.

The cloud shadows are sharpest where the specular reflection of sunshine from the sea is brightest.

## 3.5 Cold Lows

Lows are most commonly produced by convergence at fronts, but in some no fronts are prominent and often none are detectable. In such cases the upper air is colder and the convection from below up into it maintains a heat source there by the condensation of rain. Cold air lows occur in all latitudes where there is a warm sea and may even persist for some time over land if the supply of moist enough air at low levels is maintained from the sea. (See also 14.4).

In higher latitudes the sky soon becomes overcast with anvil cloud, but in lower latitudes where the upper air is drier the shower clouds are more easily seen individually.

The general characteristics of Trade Wind cumulus and cumulus over many other oceanic areas, when showers are not present are:—

(i) Convection from the sea up to cloud base does not contain upcurrents comparable in strength with those within the cloud.

(ii) The clouds are castellatus (see 1.4).

(iii) The air above cloud base is subsiding, and the drying caused thereby is counteracted by the evaporation of the cumulus into it.

(iv) The cumulus spread out as water anvils at stable levels and whether the anvils persist or not depends on the rate of warming by subsidence.

**3.5.1** In this picture cloud is beginning to accumulate at stable levels, while the tops are protruding as castellatus to greater heights, but these are evaporating very quickly. As the air moves to yet warmer sea the convection will penetrate the stable layer much more.

**3.5.2** In the Caribbean several stages of trade wind cumulus development are found. Here they have grown larger than in 3.5.1 and the anvils in the left of the picture are becoming glaciated: low down on the right are unevaporated water anvils.

The small cumulus are still typical—castellatus inclined by the wind shear. The strong thermal on the right has pileus over it, and halfway up this cloud is evidence of a second stable layer. The cirrus in the distance may be from the tops of yet larger cumulus, but this is not necessarily so.

**3.5.3** These trade wind cumulus have extensive glaciated anvils: the small clouds below are still castellatus. There are fragments of water cloud on some of the anvil edges, for example close to the centre of the picture, and the anvil to the left of it has fallout from its underside.

**3.5.1**

**3.5.2**

**3.5.3**

**3.5.4** By contrast with 3.5.1–3, in a typical cold low over Britain there are few small low level castellatus and the sky is almost overcast with glaciated anvil cloud. Haloes are not often seen in this type of ice cloud because of the falling streaks of ice crystals and the very variable density of the sheet.

The showers sometimes continue for several days and the continuous loss of heat from the anvil canopy plays an important part in maintaining the instability. At the upper levels castellatus may predominate at some stages, and this is seen in 1.4.8.

**3.6.1** This large storm in Minnesota shows on the right the newly growing cumulus, and beyond it nearer the centre of the picture, the rain, and in the centre the front edge of the cloud where it is first lifted by the advancing downdraft front (see 3.3). Above is overhanging anvil carried forward by the stronger upper wind, and most of its lower surface has mamma falling from it.

The mamma may be composed of cloud particles which have grown large enough to fall out, or they may be masses of cloud which have become colder than the clear air below as a result of sinking motion. Cloudy air sinks at the wet- and clear air at the dry-adiabatic lapse rate, which at these altitudes would differ by about 0.3°C per 100 m.

## 3.6 Mamma

Mamma are composed mainly of cloud particles falling from an anvil cloud or of downward thermals produced as a result of the subsidence of an anvil cloud. Mamma may also occur on snow or rain falling below cloud base (2.1.2) or on streaks of falling ice crystals (4.3.2, 4.3.8).

**3.6.2** This is a view of a storm receding eastwards over the plains of Colorado in the evening. The sloping underside of the anvil is illuminated by the sunshine. Together with mamma there are also fragments of cloud at stable levels in the air behind.

**3.6.4** In this case the mamma are less regular in size, and not obviously arranged in rows. It is probable that fallout particles are affecting some of them or that the anvil cloud is of very variable thickness.

It is often written that mamma indicate great instability, but actually they occur at stable surfaces made slightly unstable by subsidence, and they usually appear after storms have passed their worst periods. In any case they do represent an instability quite different from that which produced the original anvil cloud. They can occur on water anvil clouds producing no rain at all, although in that case, because of the complete absence of fallout particles, they are less well developed and rather ragged.

**3.6.6** The mamma in this case are (just above the house top) composed of masses of air containing larger fallout particles. They are irregular and are falling from a completely glaciated anvil, and are illuminated by the evening sunshine.

At the edge of the anvil is a clear ring with more cirrus cloud beyond: this probably represents a ring of sinking air around the ascending motion in the middle of the anvil. (See 1.6.3).

**3.6.3** When the mamma are produced by subsidence of the anvil cloud base and the anvil is of fairly uniform thickness the mamma are of fairly regular size. The bright gaps in between the mamma also indicate this, because masses of clear air penetrate upwards into the cloud at the same time. Prandtl was the first to give this explanation and the accompanying diagram (p56) is similar to his.

**3.6.5** It is difficult to see, from a single viewpoint, how much a cloud surface is tilted. These mamma are stretched out by shearing motion along the lines, and such shear is produced by tilting of the surface, although the tilt might actually have disappeared by the time this picture was taken.

**3.6.7** When copious rain begins to fall the bottom of the rain often has the appearance of mamma. If the smaller particles emerge from the cloud first with a fall speed which is not large compared with the downdraft speed the mamma structure may remain apparent until the rain reaches the ground, and this is most likely if the fallout is snow (see 2.1.2). But often large raindrops or hail overtake the mamma and do not quickly cause the air to sink so that they are always ahead of the front of the downdraft. In that case the rain appears as rain streaks, which are beginning on the right of this picture. (See also 1.5.4).

# 4 CIRRUS

Cirrus is a cloud of fibrous appearance. This structure is due to the stretching of the cloud by wind shear or by the fall of the particles. The cloud must be composed of ice crystals which are either not evaporating or evaporating very slowly, for water droplet clouds quickly disappear when stretched in this way. Examples of cirrus in other chapters are too numerous to refer to in detail because cirrus of some sort appears when almost any cloud is glaciated. The identification of ice cloud is often helped by the optical phenomena described in Chapter 13 but the difficulty does not arise when the cloud is clearly of fibrous appearance and is not subject to rapid evaporation. Much cirrus is of artificial origin in condensation trails, and examples of this are found in Chapter 11.

We are not concerned in this chapter with ice cloud that is not of fibrous appearance.

## 4.1 Fibrous patches

Many of these clouds remain with almost no change for an hour or two, with little movement across the sky. The shearing motions in such cases need to be of only small magnitude to stretch the cloud over many miles. Often the difference in height between the ends of a piece of cirrus is produced by large particles falling to form the lower part of the cloud; but if the air into which they fall is not quite saturated for ice they may partially evaporate and therefore fall more slowly. On the other hand the fall speed is very occasionally increased when the particles fall into air which is between its condensation level and ice evaporation level (i.e. it is supersaturated for ice but unsaturated for water).

**4.1.1**

**4.1.1** The patch of cirrus shows, on the right particularly, the effect of small wind shear in stretching the cloud elements when they are already extended vertically by the fall of the particles. On the leftmost protuberance at the bottom of the patch there are mamma, which are probably evidence of downward thermals produced by the evaporation of the falling particles into unsaturated air.

The precise origin of these patches is uncertain because, when the development is so slow it is not often that the whole life of a patch can be observed. In this case there are waveclouds in the distance, and if these become partially glaciated cirrus patches may be formed from them. They may often be the last residue of evaporating clouds of a warm front carried by the jet stream out ahead of a frontal cyclone.

The texture is clearly and obviously contrasted with the cloud beneath it, which is castellatus not very well developed, growing out of a wave cloud. The cirrus has few sharp shadows because the spacing of the ice particles is such as to cause much of the light to be transmitted. Cirrus clouds are seen mainly by the specular reflection from crystal surfaces, and are visible because of their whiteness in a blue sky, even when they are very tenuous. (Infra-red photo.)

**4.1.2** was photographed in Israel far from any weather system.

**4.1.3** Shows the effect of stretching by shear predominating over that of falling of the particles. It was taken in California.

**4.1.4** photographed over London, is the eastern edge of a warm front cloud system where the cloud is steadily thinning but thicker cloud advancing. In this case the system was not very vigorous and the shear direction is not shown by falling streaks, nor is the jet stream direction shown by cloud lines on a larger scale.

**4.1.5** is the cirrus of a wide summer warm sector over London in which the cover of low cloud has been broken up in the afternoon into small cumulus. The cirrus of tropical air in higher latitudes takes on the most beautiful shapes and often displays many optical phenomena.

## 4.2 Extensive cirrus

Most extensive cirrus sheets are frontal or formed as widespread anvil cloud which remains when the cumulus has evaporated. It also forms in areas where convergence is occurring at higher levels but not in the bottom few thousand metres, so that it is not associated with rain clouds. Or it occurs where convergence is into a summer heat low where the cumulus is sparse, but the ascent at high levels is sufficient to produce clouds. When it is produced slowly in this way it is difficult to know whether it is formed directly by growth of particles frozen from the start or whether, as is more probable, it is first formed at water saturation and quickly frozen because of the low temperature (below about −38°C).

If it is formed at water saturation it is probable that the particles will grow by sublimation and acquire fall speeds large enough to give texture to the cloud.

**4.2.1** On the left are patches of leaf-like cirrus which suggest a falling motion concentrated along a spine as in 4.3.5. On the right is part of a long band of cirrus with striations across its length. This is very typical of orographic cirrus (5.4) and of jet streams which often contain much orographic cirrus. This picture was taken looking towards the dawn at Milford Haven in South Wales, and the cirrus probably originated over the Wicklow Mountains of South East Ireland.

**4.2.2** This is a good example of convergence in the upper air only. This cirrus thickened over many hours and produced copious snow which fell out in large mamma. It was like anvil residue but occurred in the absence of low cloud. The snow evaporated in the air below dried by subsidence. This was seen over Exmoor in early June.

**4.2.3** The fibres are drawn out in two directions in this cirrus (sometimes called cirrostratus, a fibrous layer) ahead of a warm front. Looking towards the west we see the upper level streaks pointing towards the NE, indicating that the warm air is flowing outwards above the cold air ahead of the front. Beneath the striations are drawn out in the direction of the thermal wind, which is the same as the direction of the jet stream; the wind and shear are from the NW with the cold air on the left when facing downstream (see also 4.2.6). The cloud thickens towards the west. The foreground is blurred because the picture was taken from a moving train.

**4.2.4** As the warm air advances at a warm front the sun is gradually obscured by thickening ice cloud in which haloes (see 13.2.3) may be seen and in which contrails are common and dense. The cloud is called cirrostratus or altostratus and often thickens into nimbostratus (which is not illustrated because it produces a uniformly grey sky with rain or snow).

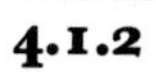
4.1.2

4.1.3

4.1.4

4.1.5

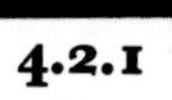
4.2.1

4.2.2

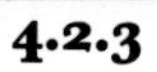
4.2.3

4.2.4

**4.2.5** This shows the classical view of a distant warm front with cirrostratus thickening above scant cloud below.

The shadow of the aircraft's contrail is seen at the bottom of the picture.

**4.2.6** Extensive sheets of cirrus are often situated just below the tropopause, and may have originated in frontal systems in which the lower clouds have been evaporated by subsidence. In this example over the Mediterranean west of Rhodes the cirrus appears as dark cloud when seen with cumulus in the background, which is not usual. The tropopause is seen as a sharp horizon.

**4.2.7** A jet stream is often characterised by a fairly sharp edge to the cirrus cloud, with striations lying at right angles to the edge. The cloud moves with the jet stream whose motion is almost parallel to the edge, but with a slight component of motion across it. In the northern hemisphere the cold air is on the left of the jet stream, and in this picture, taken at Seattle, in which the cloud edge lay SW–NE, the motion was from top left to bottom right.

There are two components in the motion at right angles to the cloud edge. The first is the general movement across the ground of the front associated with the jet stream. In the case of a cold front the cloud recedes as the clearance advances with a speed that is usually much smaller than the jet stream strength. The second component is due to the overturning of the air: the cold air sinks under the warm air and the energy released accelerates the jet stream. The warm air, in which the cloud is formed, moves outwards over the cold air, and any falling cirrus particles make trails pointing inwards towards the warm air.

This was taken near a jet stream entrance where the overturning motion was quite strong; but the front moved away and it was far to the east that the force of the developing winds was experienced.

## 4.3 Falling Cirrus

**4.3.i** shows crudely the idea that when some of the particles become glaciated in a patch of water cloud, such as an isolated castellatus thermal, they grow rapidly in the supersaturated environment of the unfrozen water droplets (see chapter 2 and 7.4) and soon fall out, ultimately arriving at a level at which they evaporate. While they fall wind shear stretches the cloud into long streaks.

This simple situation may be modified by the complete and almost instantaneous glaciation of the parent cloud, by increased or decreased rates of fall due to condensation or evaporation or by variations in the strength of the shear at different levels.

**4.3.1** was taken in mid afternoon looking southwards near Bardai (Tibesti, central Sahara). The parent clouds have quickly evaporated, and either the shear increases or, more probably, the fall speed decreases towards the bottom of the streaks where they become almost horizontal. The main streak is about 7 km. in length. The particles fall at speeds up to about 1 metre per second.

**4.3.2** is a view from almost vertically beneath two almost completely glaciated castellatus clouds from which a copious stream of ice crystals is falling. Mamma can be seen on the lower extremities of the streaks where the crystals are falling into drier air and evaporating rapidly. The contrast has been intensified by the use of a polarising filter.

**4.3.3** Isolated thermals are common, but do not often show the ring-shaped structure of the motion because the upcurrent in the middle is usually filled with cloud. The thermal in the centre of this picture is evaporating, having also produced a trail of ice crystals which are beginning to be drawn out by shear. This view is vertically upwards.

**4.3.4** Just as mamma can be formed on the base of an anvil subjected to subsidence, it can also be produced when the particles of a newly glaciated sheet of cloud begin to fall. The downdrafts produced by the falling particles which add their weight to the air into which they fall and which they may also cool by some evaporation, cause later crystals to fall down the same tracks, and not through the updrafts in between. Often the updrafts produce clear holes in a sheet of cloud. When illuminated obliquely by the setting sun the lower tips clouds of falling particles are often brightly illuminated while the crystals just above, and perhaps the remains of the parent cloud also, remain in shadow and can scarcely be seen.

4.2.5

4.2.7

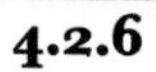

4.2.6

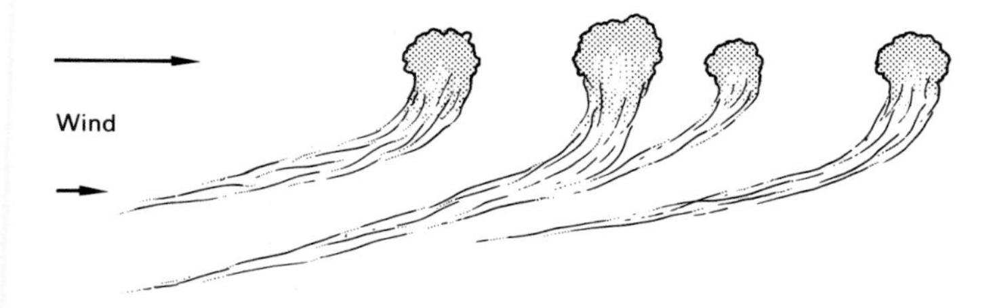


4.3.i

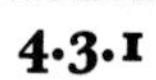

4.3.1

4.3.2

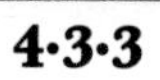

4.3.3

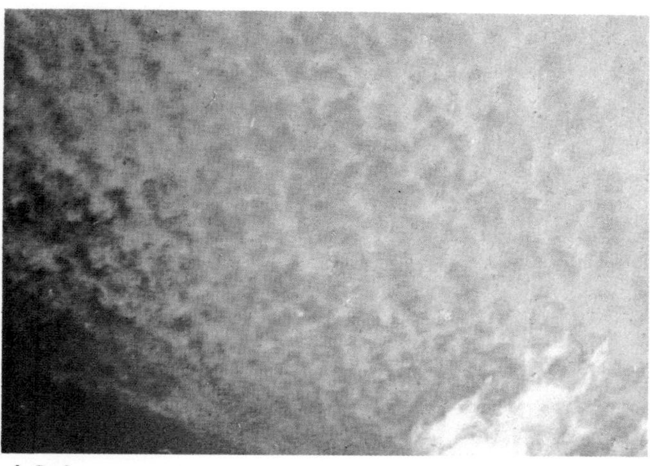

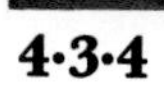

4.3.4

**4.3.5** Fallstreak holes are produced in a thin layer of supercooled cloud when glaciation begins at isolated points, from which it spreads outwards.

The crystals fall so as to form a horn-shaped surface with the first crystals falling from the centre at the bottom of a streak which widens upwards.

The glaciation is thought to spread through a cloud by means of the small splinters of ice ejected from a droplet which has grown a shell of ice on its outside first and is then shattered when the water inside also freezes and expands. The splinters cause the freezing of other droplets with which they make contact.

It is usually uncertain how the original glaciation is localised at a few points. Most probably it is initiated by the passage of an aircraft, for at low temperatures droplets become frozen, and probably shattered, on contact with the cold aircraft surface. There may also be low pressure regions around the aircraft where the air is cooled to −40°C or below long enough for glaciation of the cloud droplets to occur spontaneously.

Another possibility is that the holes were started by crystals falling from isolated fall streaks above the layer.

**4.3.6** Radar echoes are obtained predominantly from the larger particles such as snow and rain. This radar picture which is a vertical section of the air, shows the fall streak from snow-generating cells at about 20,000 ft. stretched out over 30 to 50 miles near Montreal.

**4.3.7** If a continuous picture is made of the echo received in a vertically pointing radar, a cloud or rain system passing overhead will be depicted as a vertical plane section of the atmosphere. The uppermost snow producing cells give rise to ordinary fall streaks in among much smaller crystals which are not seen by the radar. At 7,000 ft. where all the crystals melt on arriving at the 0°C level the echo is greatly intensified because wet crystals are much better radar reflectors than dry ones. Beneath the bright band at the melting level the intensity decreases because the particles become spherical and quickly reach a higher fall of speed which reduces the concentration of particles in the air. The lower streaks are of rain.

**4.3.8** The falling snow seen on this radar time section is evaporating in drier air and producing mamma which entrain other crystals as in 4.3.4. Some of these columns appear thinner than they really are because their widest part did not pass immediately over the radar. Although no snow has reached the ground at the radar station during the period of observation, some of these snow streaks did so after about an hour.

The time section does not give a correct picture of the cloud structure at any instant because the wind is stronger at the higher levels in this case. These cases show typical warm front snow at Montreal, and snow reached the ground at the station about four hours after the first falling snow was seen overhead.

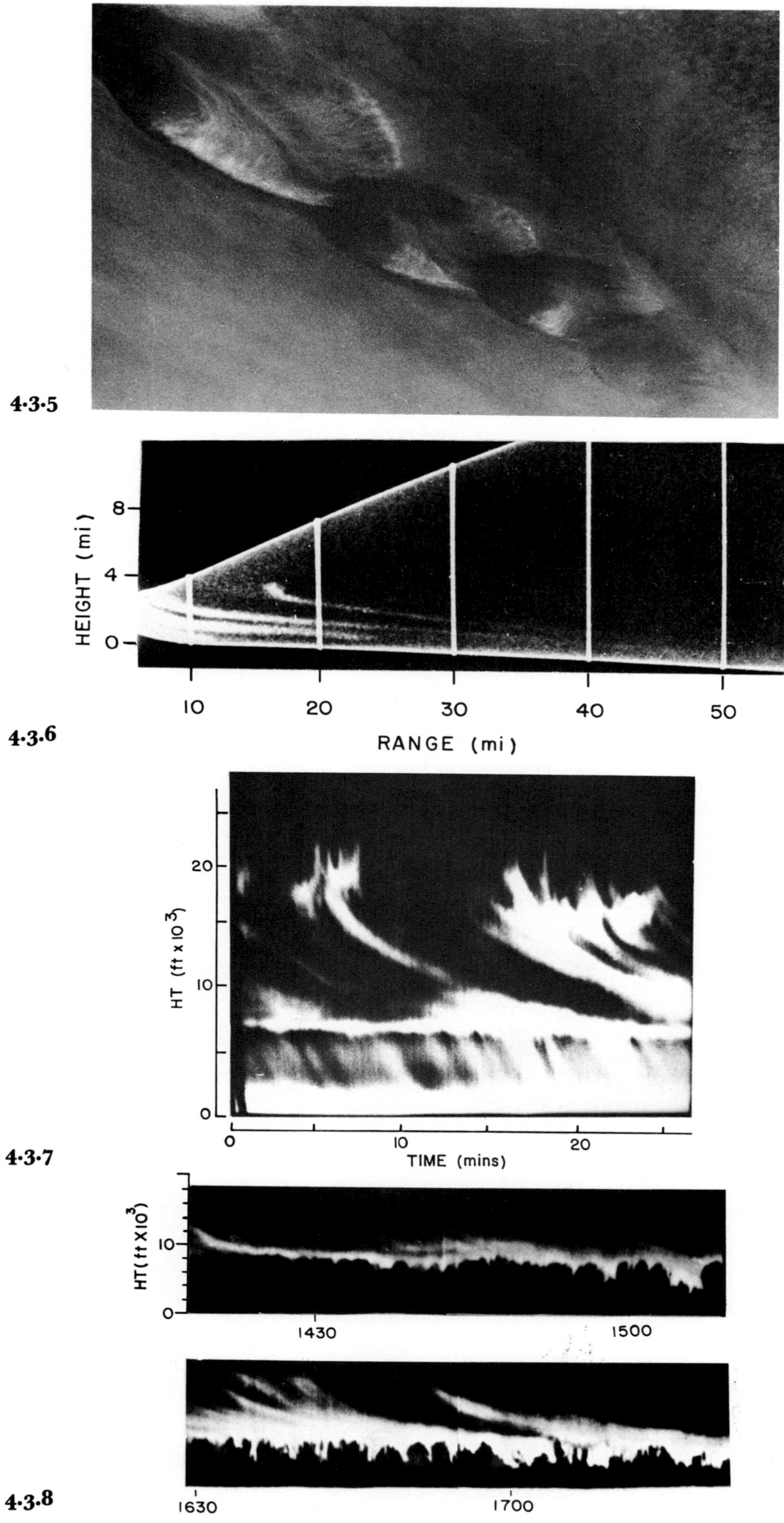
4·3·5
HEIGHT (mi)
8
4
0
10
20
30
40
50
RANGE (mi)
4·3·6
HT (ft x 10³)
20
10
0
0
10
20
TIME (mins)
4·3·7
HT(ft X10³)
10
0
1430
1500
1630
1700
4·3·8

# 5 WAVE CLOUDS

The most significant wave clouds are those found in the crests of standing waves. Although several early writers on clouds showed that they understood them correctly they were not specified in the accepted international coding, presumably because no use for them was seen in preparing weather chart analyses and prognostications. It was left to glider pilots to explore and exploit them while other aviators often courted catastrophe by ignorance of them. They are very common; even at considerable heights over small hills. When the wave pattern is not steady their behaviour becomes erratic and is not as well understood because the mathematical theory is only tractable for steady motion. Even then they are fairly easy to recognise because the clouds as a whole move at a different speed from the wind which carries the cloud elements with it.

The conditions most favourable for waves in the troposphere are an increase of wind from the middle troposphere up into the stratosphere with a decrease in static stability from the lower to the upper troposphere. If the stability of the lower layers is concentrated at an inversion, for example at the upper surface of a layer of stratocumulus, the wave amplitude is almost invariably a maximum at that level.

Clearly if the conditions just specified were found lower down the waves would occur in the lowest layers containing the more stable air and/or the lighter wind, and would decrease in amplitude up through the layers of smaller stability and/or stronger wind. The theories of standing waves are fairly precise in their conclusions about the air flow for tropospheric waves, but are not yet properly developed for waves in the stratosphere nor for cases in which the wind component in the direction of the wind near the ground drops to zero at some higher level. There is one obvious requirement for waves in the stratosphere, namely that there should be a moderate wind component across the mountain at all heights up to at least a little above any wave cloud.

## 5.1 Mountain Waves

Waves in the immediate neighbourhood of mountains, which means, for practical purposes, more or less above the mountains, are called mountain waves, to distinguish them from the lee waves which are found in some cases up to 200–300 km. downwind of the mountain although we are generally concerned with the 20 km. closest to the mountain.

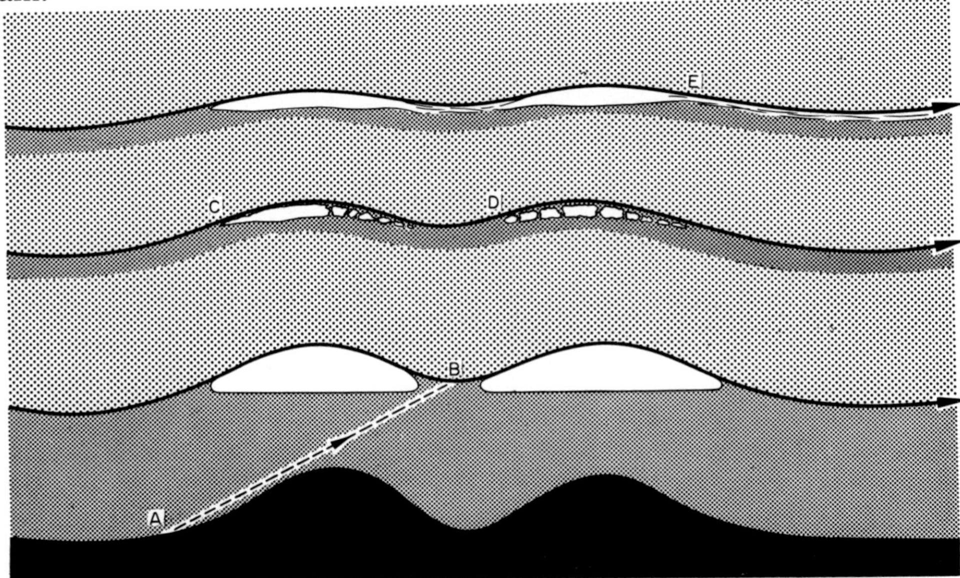


5.1.i

**5.1.i** The streamlines at higher levels do not usually have quite the same shape as the ground profile, but they are often rather like it. To illustrate the cloud mechanisms they are drawn that way here over a double hump profile. The layer below the lowest streamline is assumed to be well mixed, probably by convection with the streamline drawn to represent the bottom of the sub-cloud stable layer, so that the air above it is stably stratified. Cloud is formed in this air with a flat base and arched top when the waves lift it above the condensation level. From A the smooth top of the second wave can be seen through the wave hole at B (see 5.2).

A second moist layer is envisaged at the level of the next marked streamline, but it is assumed that the layer is unstable when cloud forms in it. Therefore as the air passes through the wave it breaks up into lumps because of convection. The lumpiness will reappear in the second wave at D although the edge where the condensation first occurred at C was smooth.

The uppermost moist layer drawn here is assumed to have a temperature close to or below −40°C so that at least some of the particles freeze and do not evaporate between the waves or downwind of them at E. This theme is developed further in 5.4 and 5.7.i.

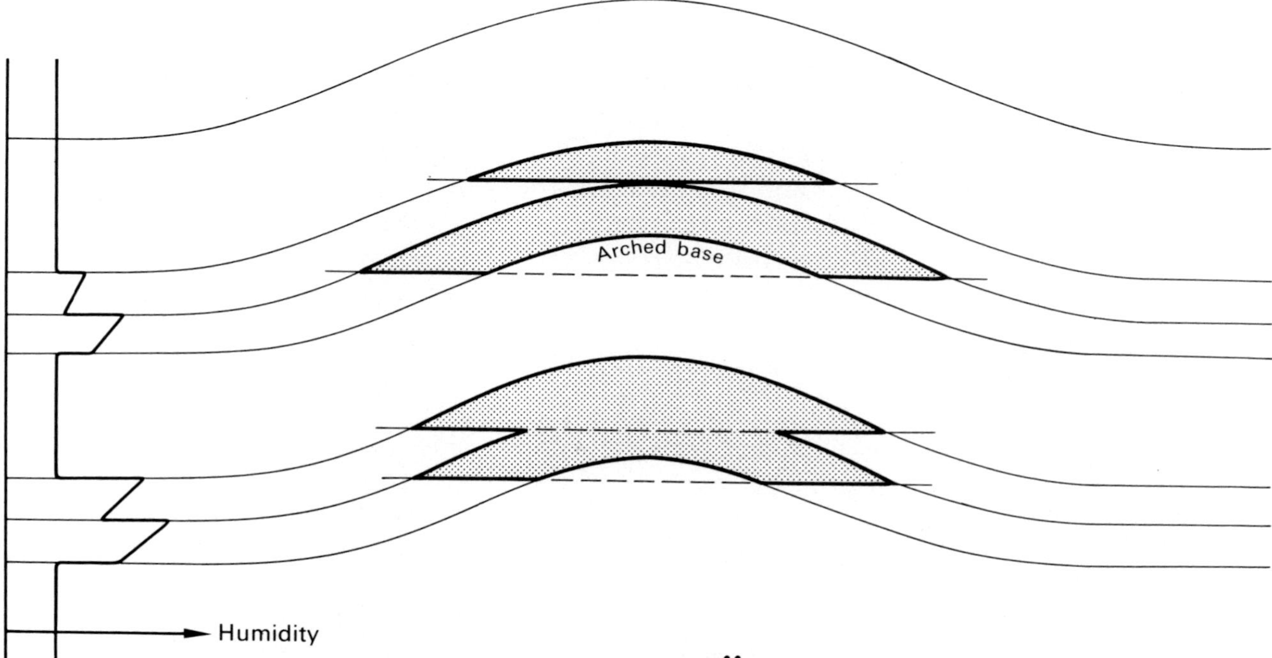

**5·1·ii** Some wave clouds appear to be composed of several layers. This is because of a layered structure of the humidity. On the left is an imagined profile of relative humidity, each layer with its own condensation level. The shape of a cloud in a mountain wave is shown on the right. Clearly the base of the wave cloud could be arched if the layer lifted into the base of a wave cloud were dry enough.

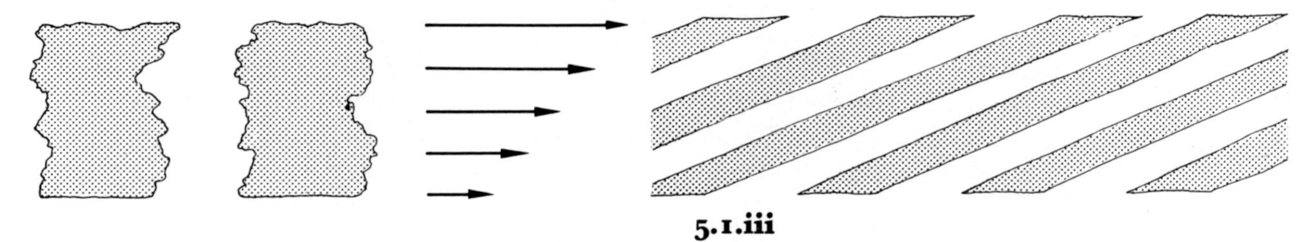

**5·1·iii**

**5·1·iii** It is not usually obvious how the humidity structure becomes laminated in any particular case in which layers appear. But it is known that the moisture carried upwards is transported either by cumulus clouds or by inclined flow up frontal surfaces. In the latter case there is a large amount of shear present so that moist tongues are often drawn out into thin sheets.

When the air is moistened by large cumulus towers these are often sheared over so that after only a few hours the top of a moistened tower of air may be hundreds of miles from the bottom and the compact mass of air stretched out into a very thin long sheet, which soon becomes almost horizontal with other sheets above and below it. This mechanism must operate because the air in between cloud towers is stable so that all mixing motions cease when the clouds evaporate (see 1.1.3). But the effect is usually swamped by subsequent towers rising up through the now almost horizontal layers. Occasionally however this is not so and the laminations are preserved.

**5.1.1** Completely smooth tops to wave clouds are rare because of the formation of billows and corrugations (see 6.1.iii) on them. There are several systems of billows of different wavelength on this wave cloud seen in stereo over north eastern Oregon. In the distance are more wave clouds of the same moist layer. The right edge of the cloud has a smoothness typical of the upwind edge of the first wave (see fig. 5.1.i).

**5.1.2** During flights above complete layers of cloud the only hint of the nature of the ground below is obtained from the cloud forms. These long shadows are cast by the low sun shining from the north across Baffin Bay in early July. This tenuous wave cloud together with the gaps and raggedness of the many layers below indicated that the aircraft was crossing the west coast of Greenland on a flight from Seattle to London. In a stereo picture of this kind the cloud can be placed relative to the others better than from the aircraft in flight because the relative motion of transparent layers is not always easy to interpret at sight.

**5.1.3** When the sky is otherwise clear the mountains of Greenland produce wave clouds most beautiful in their simplicity. These were often called whale-back clouds by the great whaling captains of the nineteenth century who observed them over distant coasts. It is notable that although the ground is snow-covered these smooth clouds are not glaciated.

**5.1.4** Thin sheets of cloud such as this reveal wave motion particularly well because of the holes in them. They are at the level of maximum amplitude of the wave motion. The distant edge lies close to the western coast of Pembrokeshire in South Wales and the wind is from the north west in late afternoon in August. Note the billows, which are usually located on the cloud top and are the last part to evaporate in the wave troughs.

**5.1.5** The wind and stability structure are very often favourable for the formation of waves ahead of a warm front. The sun is becoming obscured by the thickening layers of ice cloud at the highest levels; beneath, the first clouds to appear in middle levels will be in wave crests, and in many localities these are regarded as more reliable portents of coming rain than the well known 22° halo (13.3). This view is taken looking south westwards at Peel in the Isle of Man. The wave cloud in the top right corner and in the distance beyond the castle are quite typical warm front wave clouds. The upper cloud has a diffuse base because of the particles falling from it. It is usually called altostratus (a layer not produced by direct ground influence) but it is in fact a completely glaciated cloud (comparable in structure with 4.2.2, but produced by a quite different cause).

**5.1.6** The top of this cloud is wrinkled (top right corner) possibly due to billow formation, while the lower layers are very stable with laminations of humidity. In the distance the cloud is cumuliform over the snow-free mountains in Southern Greenland. It must be remembered that in summer the days are very long and the sun rises to an elevation of over 50° in midsummer at midday, so that the heating may be very appreciable.

5.1.1

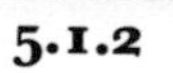
5.1.2

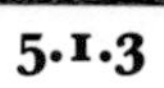
5.1.3

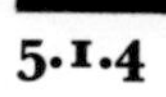
5.1.4

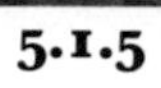
5.1.5

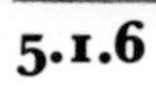
5.1.6

**5.1.7** A laminated wave cloud is often called a "pile of plates". This example was observed from the air over an isolated oceanic island near S. Georgia. A pileus-like wave is seen in the distance over the crest of a low cloud. The "plates" appear to be hollowed out on the under side so that each rim is illuminated by the low sun which is beyond the low cloud on the right.

**5.1.8** The bottom of this unusual "pile of plates" was only about two thousand feet above the ground. It was seen in August in southern Alberta. The laminations indicate considerable stability, and clearly the condensation level of the lower layers was lower. When a picture of this cloud was first published it was described as a 'vortex cloud', but as far as the air motion is concerned it is a fairly ordinary wave cloud.

**5.1.9** The white clouds aloft are laminated wave clouds over Harlech on the west coast of Wales in an east wind which carried pollution from central England over the Welsh mountains. Because of the great stability the pollution remained confined to the lowest 600 metres or so, and presents here a fairly sharply topped haze layer. It is of special note that the clouds appeared not over the higher mountains inland but where the ground descended steeply to the sea. They were probably at a height of about 3,000 metres.

**5.1.10** This is part of a wave cloud formed in a very light wind at Fort Collins, Colorado, close to the Rocky Mountains. The left edge is typical of a smoothly formed wave cloud, with streaks along the direction of the wind. On the right it is becoming thicker and also unstable so that by convection it breaks into lumps.

On this occasion the waves were not steady. Convection was beginning to alter the flow pattern in mid-morning and the air motion close to the mountains was changing particularly rapidly as anabatic flow and separation from the mountain crests began. These waves appeared over small hills of the low ground before convection from that ground produced its own cumulus.

## 5.2 Wave Holes

Where the air descends between waves a gap in the cloud is often formed, particularly in thin layers (see also 6.1.2).

**5.2.1** Wave clouds can be formed in among cumulus: this is possible because the air between cumulus clouds is stably stratified, but the amplitude of the waves is usually smaller than over the same hills when no cumulus are present. The edges of the cloud around this wave hole have 'fingers' extending from them along the wind direction: these corrugations, described in 6.1.2, occur when the streamlines have a large curvature, and this happens mainly in waves of short length. This gap was over a small valley near Cader Idris (North Wales).

**5.2.3** The wind descends over the Mawddach valley in Wales leaving a hole in the cloud through which the smooth top of warm sector stratocumulus can be seen.

**5.2.4** The wind is from the west (right) over the northern part of Corsica. The cloud gap is in the lee of the coast, and on the left is the lee wave cloud, with a higher level lee wave cloud above it.

**5.2.2** The lake is Ullswater and the rainbow was visible for about 45 minutes without interruption, in spite of a fairly strong wind, because of a more or less permanent gap in the clouds over the lake. The bow was formed in drizzle which, because of its slow fall speed, reached the ground under the gap. This is a fairly common phenomenon, but the bows are only visible from one direction. For example, there is often an almost permanent rainbow in the afternoon to the north of Honolulu where the trade wind cloud produces drizzle over the mountains, and as the air descends the lee slope the cloud evaporates and the drizzle falls in sunshine.

Note that the rain shines brightly inside the arc which represents the minimum angle from the antisolar point (observer's shadow) through which the rays can be deviated, the red being deviated least. The secondary bow, which represents the extreme deviation of a ray twice reflected inside a drop, can be seen faintly outside (see 13.2).

5.1.7

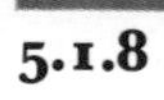

5.1.8

5.1.9

5.1.10

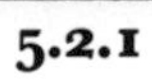

5.2.1

5.2.2

5.2.3

5.2.4

## 5.3 Lee Waves

In airstreams with the structure already described above as suitable for mountain wave formation, trains of lee waves are often formed. The amplitude of these waves is very dependent upon the shape of the mountain, while the wavelength depends entirely on the characteristics of the airstream. The waves are those of such a length that they travel with a speed exactly equal and opposite to the wind, and so they remain stationary relative to the ground. They occur often in clear air, but they are also quite common in thin layers of cloud because the loss of heat by radiation from the cloud top usually forms a very stable layer beneath the clear air above, at which the wave amplitude is a maximum.

Wavelengths are shorter in more stable airstreams with lower wind speeds.

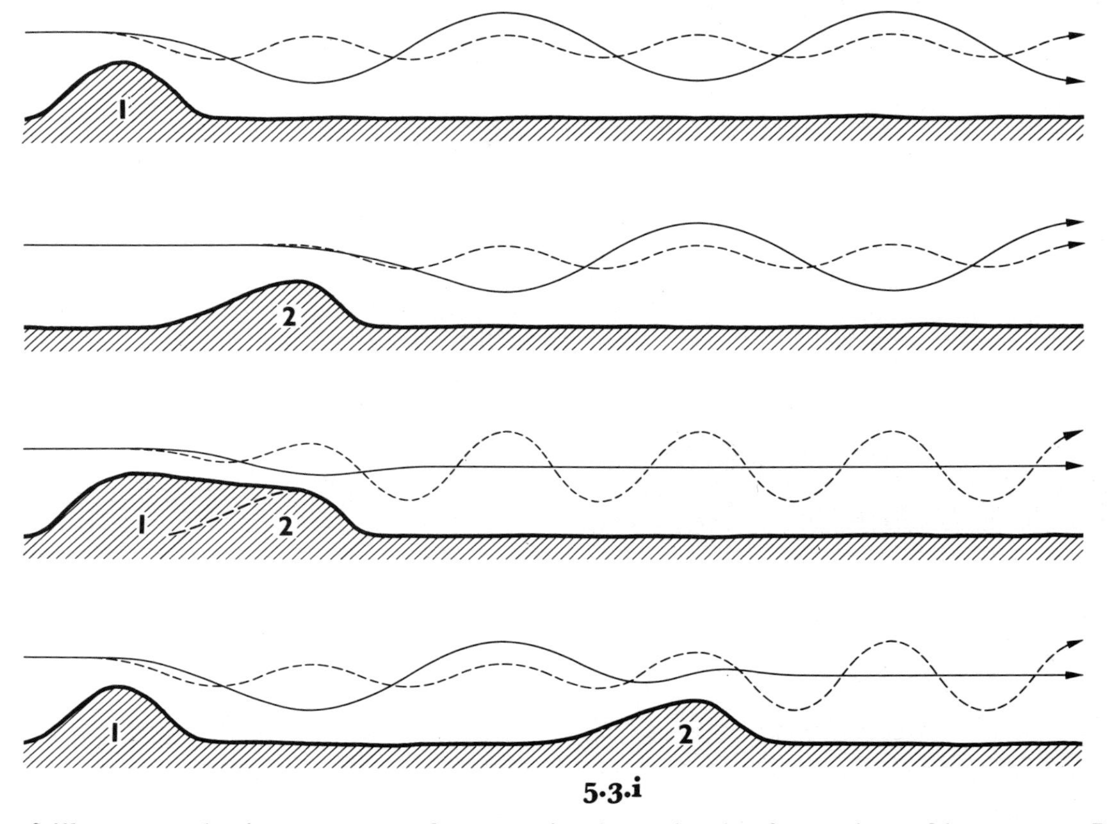


**5·3·i**

**5·3·i** illustrates the importance of mountain shape in the formation of lee waves. In the first place the mountain must have a width which is of the same order of magnitude as the wavelength. If it is too wide or too narrow the wave amplitude is very small. The continuous and dotted lines represent waves in two different streams set up by flow over the mountain ridges 1 and 2 separately. That with the shorter wavelength being in a slower moving and/or more stable airstream.

A mountain whose profile is the sum of ridges 1 and 2 will produce a wave which is the sum of the two separate wave trains. In the first airstream the two trains are cancelled out, in the second they reinforce each other.

When the ridges are separate we see that in the first airstream a fairly large wave now appears in front of ridge 2 but the lee waves behind it are cancelled out. In the second airstream the lee waves are again reinforced. A small difference in the distance between the ridges could cause the lee waves of the first airstream to be amplified and those of the second to be cancelled out.

As a consequence of this very sensitive relation between total mountain shape and lee wave size, no simple rule can be given for probable wave amplitude, and in a region of complicated topography large waves may be found distributed among regions of almost horizontal airflow.

An isolated mountain produces behind it a widening wave pattern, and in a region with peaks rather than ridges the waves tend to form across the airstream by the reinforcement of overlapping wave trains from peaks. Downwind of parallel ridges the lee waves are parallel to the ridges, and not necessarily at right angles to the wind.

**5·3·ii** is a sketch to show the scheme of pictures 5.3.1 and 2. There is a 'helm cloud' over the mountain range with a series of helm-bars, or roll clouds, parallel to it on the lee side. There is a strong wind down the mountain slope and almost calm regions beneath the cloud bars. The roll clouds often appear to be tumbling over continuously because of the stronger wind at the top of them. Sometimes the wind is reversed at the ground beneath them (see figure 5.7.i).

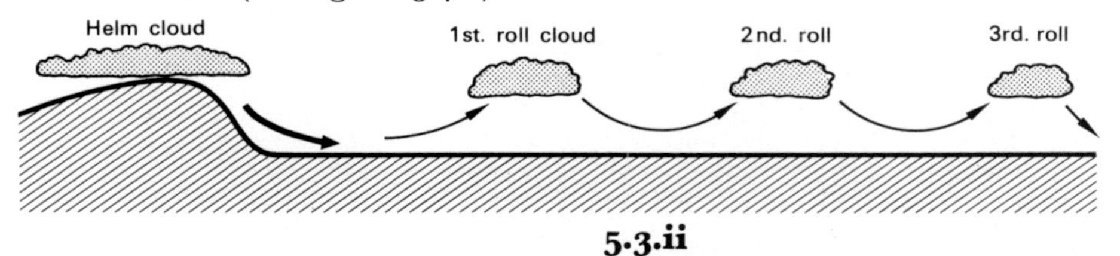


**5·3·ii**

**5·3·1** Shows a helm cloud and the first bar cloud seen over Mt. Hira from a train during a typhoon. The wind is blowing from the mountain towards the observer. The near edge of the helm cloud is evaporating and the fragments on the far edge of the bar cloud are growing and moving into it.

**5·3·1**

**5·3·2** Roll clouds in the lee of Mt. Suzaka showing the ragged evaporating edge of a bar overhead. There are clear gaps visible between the second and third bars and the fourth is partly hidden by the third. The view is from a train and telephone wires obstructing the view have been scratched out of the picture.

**5·3·2**

**5·3·3** Waves on a haze top are best seen when the sun is very low and is shining almost along it. Those seen here were formed at about 400 metres on the west side of the Isle of Man in an east wind. There were several fires, from one of which the flames can be seen, in the heather, and the smoke from these spread up to the haze top and made it more sharply defined. In this case the sun had already set and the waves are visible where the distant glow of the sky is seen through the wrinkled haze top.

**5·3·3**

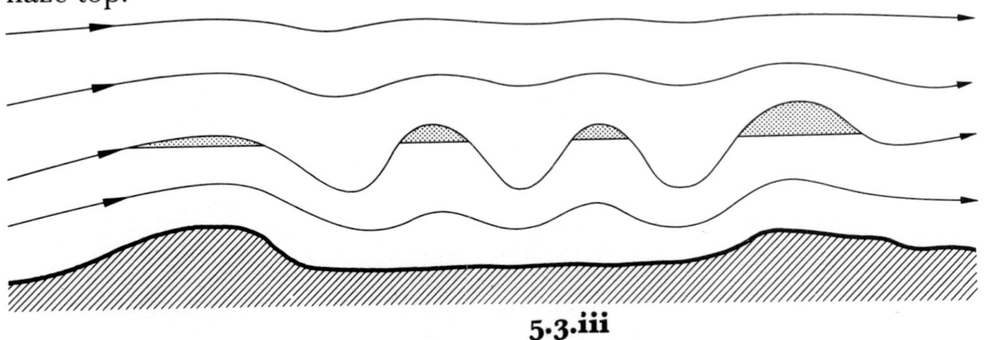

**5·3·iii**

**5·3·iii** In this case we have imagined that the third lee wave occurs over a second hill so that if clouds appear in the wave crests the fourth cloud is the largest although it is over a smaller hill than the first (see 5.3.4).

**5.3.4** The nearest cloud is in the third wave in the lee of a mountain about 4 km. away under the first (most distant) cloud. Figure 5.3.iii was drawn to illustrate this case.

When seen near to the sun clouds like these are often brightly iridescent (see 13.2.3). On account of their smooth appearance they are often thought to be at considerable heights: furthermore they do not move even in a strong wind and this enhances the illusion. These clouds were at about 1,000 metres near Towyn (North Wales).

**5.3.5** The dark bands in these lee wave clouds are where arched lee waves are viewed tangentially. The cloud there is also darker because of the low sun. The wind is towards the observer. (S. Ireland).

**5.3.6** This is a layer of cloud set in wave motion over the Adriatic Sea near Corfu. The wind is from the NE, even at the level of these clouds, which are at nearly 10,000 metres and yet remain unglaciated. The aircraft is flying about 700 metres above them and the stable layer at their top which is probably strengthened by radiation from them, has a very similar appearance to the tropopause, which was probably another 2,000 metres higher on this occasion.

**5.3.7** The Royal yacht Britannia at anchor off the coast of Greenland has its flags extended towards the coast, yet the wind is away from it. We see the evaporating lee edge of cloud as the air descends to the sea, and the characteristically smooth upper surface like the whale-back clouds of 5.1.3. In the top of the picture is the front edge of the first lee wave cloud into which the air ascends. Beneath this wave the wind is occasionally calm or reversed (See fig. 5.7.i); thus the ship faces the land but at the moment of the picture the wind is astern.

**5.3.8** The wind blows diagonally from the bottom right corner, along the cloud streets formed by convection over land (see 1.3) about the middle of the picture on the right. When the airstream crosses the coast of Lyme Bay (S. England) at which there is a fairly steep cliff, it is set into wave motion and 8 or 9 lee waves can be seen in the cloud out to sea. The rather uneven topography is probably the cause of the absence of cloud streets over the land in the left and bottom of the picture.

**5.3.9** As the air descended the east side of the hills to the south of San Francisco, into San Francisco bay, the cloud was evaporated. In this case the descent was in the form of two steps, as if there were a lee wave of rather short wavelength half way down the slope. The cloud layer extended for a great distance over the ocean with an almost flat surface: the shadows from the evening sun tend to exaggerate the rather small descent required to evaporate the thin cloud.

## 5.4 Glaciation in Waves

In figures 5.1.i and 5.7.i the uppermost wave cloud is shown with a trail of ice cloud extending downwind as far as the trail remains above the ice evaporation level. These trails are known as orographic cirrus.

**5.4.1** The whiter cloud in the centre of the sky is an unglaciated wave cloud and a trail of frozen particles extends downwind from it to the left: this trail is somewhat brighter than the more tenuous ice clouds in the rest of the sky. There is some lower cloud in the hazy distance over the northern Apennines of Italy.

**5.4.2** The low wave cloud on the left has a slightly arched base, which is rather unusual after a day of convection. The cloud also has longitudinal corrugations (see 6.1.iv). The cloud across the top is orographic cirrus carried downwind from the cloud at its right end. (This was seen near Loch Ness.)

5·3·4

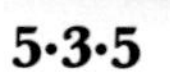

5·3·5

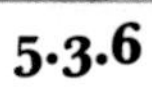

5·3·6

5·3·7

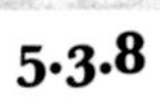

5·3·8

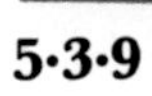

5·3·9

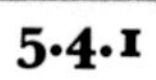

5·4·1

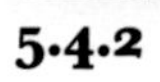

5·4·2

**5.4.3** There is a water cloud clearly discernible just above the small dark mountain on the right. Above, and also streaming from the left edge of this cloud are arched streamers of orographic cirrus. The cloud on the left is also unfrozen. These mountains of Greenland protrude from a snow field which lies like an inland sea, at several thousand metres above sea level.

**5.4.4** The wind blows from the left along the streets (see 1.3) on the western side of the Pennines near Buxton, England. High above over the High Peak is a wave cloud whose western edge appears to be glaciating immediately on formation, and is therefore probably at a temperature below −40°C. Another possibility is that such cloud is growing on ice nuclei on which cloud had formed and then evaporated in a wave over a mountain not far upwind: according to this theory these nuclei would still be in a condition such that they would grow as soon as ice saturation was reached. Normally the first cloud forms only when water saturation is reached.

**5.4.5** Iridescence is seen in clouds whose droplet size is fairly uniform at a point. The colour seen depends on the angle from the sun and the droplet size. Close to the sun rings are seen (see 13.2.2 and 3) concentric on the sun, but at greater angular distances the variations in droplet size become more important, and the colours therefore tend to follow the contours of a wave cloud in which, owing to the relatively weak upcurrent compared with a cumulus, only the more efficient nuclei grow droplets, their number density remaining fairly constant through the cloud.

Any wave cloud might, in principle, show iridescence but it is most clearly visible where the rest of the sky is dark. Mother of pearl clouds are at altitudes between 19 and 30 km. where they may be brightly illuminated by the sun after sunset at the ground, and after the sun has ceased to illuminate the troposphere where most of the blue colour of the sky originates. A relative shortage of nuclei may severely limit the number of cloud droplets and make their size rather more uniform than in tropospheric clouds.

This mother of pearl cloud, seen near Oslo, is in a NW wind so that the bottom (most distant) edge of the cloud is where the air enters it. Trailing from the eastern edge is a stream of frozen particles which do not evaporate where the unfrozen particles do. A cloud of this appearance in the troposphere could be assumed to be at a temperature slightly above −40°C, so that only some of the particles would freeze. But other evidence shows that these clouds are at between −80°C and −90°C, and it is thought that some of the particles remain unfrozen, and evaporate at the lee edge of the wave like water droplets, because of their small size (about 0·15 microns). But small size alone could not very much delay glaciation at −80°C, and it is probable that the nuclei on which the droplets form are hygroscopic, possibly $H_2SO_4$ which has been synthesised at some great height, and the freezing is mainly delayed by the high concentrations of solution.

The iridescence cannot be explained in terms of ice crystals, nor in terms of ice spheres because they would not evaporate at the lee edge and leave a trail of other particles to be carried away downwind.

### 5.5 Castellatus in Waves

Since the air might be unstable when it becomes saturated, convection from a wave cloud might extend up into the air above it in the form of castellatus (see 1.4.1, 2, and 4). This is fairly common in waves, and most castellatus seen over land in the early morning or late evening is probably formed in waves.

**5.5.1** The upper layer of ice cloud is clearly in wave motion and although over London, looks rather like the Greenland cloud of 5.4.3. Beneath it, and looking darker, are some thin unfrozen wave clouds some of which have castellatus tops.

**5.5.2** The wave motion in this case is very slight and the wind is almost along the apparent lines of the waves from the left. Where the amplitude is just sufficient for the wave cloud top to be warm enough relative to the air above for it to sprout towers upwards, castellatus is formed: the shallower waves, as in 5.5.1, do not produce towers.

5·4·3

5·4·4

5·4·5

5·5·1

5·5·2

## 5.6 Unsteady Waves

**5.6.1** Waves are unsteady from a variety of causes. Castellatus exemplify the unsteady sprouting of wave cloud tops, but these show a more ragged top whose cause is uncertain. There is some similarity to 5.7.6 which suggests that the air above the cloud is being lifted—the ragged cloud is a kind of scud in the air above. These 'hedgehog clouds' are rare and shortlived and are probably caused by large amplitude unsteady waves thrown up like a splashing wave in a choppy sea. Their essential feature is that they are not smooth, like pileus, which never shows a ragged top of this kind.

**5.6.2** In the early morning when pools of cold air in valleys become stirred up, the flow higher up becomes more wavy as the ground contours are more nearly followed by the surface wind. This picture was taken on an early July morning at Camphill (near Sheffield) as patches of morning mist and guttation on the grass are dried out to form scud-like cumulus which evaporates as it rises into the air above. The complicated wave pattern in the thin layer at about 1,200 metres changes rapidly as the wind begins to follow the ground slopes.

**5.6.i** The streamlines show how even in the absence of lee waves the flow from a plateau may be accompanied by a disappearance of cloud at some levels, or the appearance at another. The latter was the most remarkable feature of flow in the lee of a mountain that was noticed by glider pilots first exploring the airflow in the lee of the Alps.

**5.6.3** A common characteristic of unsteady waves is the appearance of wave clouds of much smaller wavelength (see also 5.6.6). On the right lie the Sierra Nevada (California), on the left the Inyo Mountains, and the wind is from the west (right). We look along the west edge of the clouds in the first lee wave (W in fig. 5.6.i); there is also a step up in the streamlines at the cloud level as the flow at the ground steps down into the Owens Valley (see 5.7.1 for aerial view of the topography). In this case the flow is unsteady and contains a large number of small pileus-like wave clouds which last for only a few minutes: their bases are ragged rather like the right hand wave in 5.6.6.

**5.6.ii** is a diagram of the unsteady motion seen in 5.6.4 and 5, which are views from the mountain top towards the sun. Where the flow rejoined the sea surface (J) occasionally the wind gusts caused a darkening of the almost calm surface by new small waves. The separation (S) was located at the top of the steep lee slope.

**5.6.4** This view was taken during the afternoon before 5.3.3 on the Cronk ny Irree Laa in the Isle of Man. The wind is blowing down a steep lee slope up which there is an anabatic flow in the April sunshine. This breaks away from the salient edge in the form of a series of thermals which produce a rapid pulsation in the inversion which traps them (see fig. 5.6.ii). At the top of the picture are three quite separate upward surges. The sun, seen faintly through the cloud, illuminates the top edge tangentially. Nearer the camera are other thermals on their way up to the inversion. 5.6.5 shows the view a few moments later.

5.6.1 a

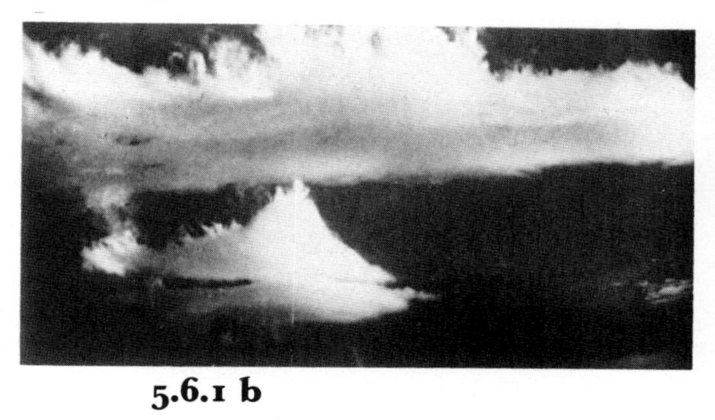

5.6.1 b

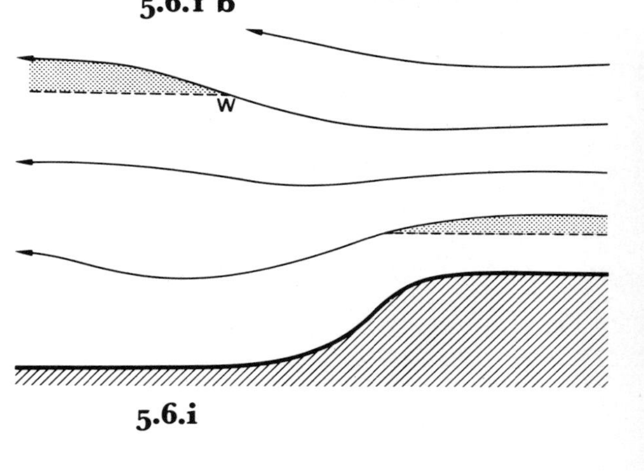


5.6.i

5.6.2

5.6.3

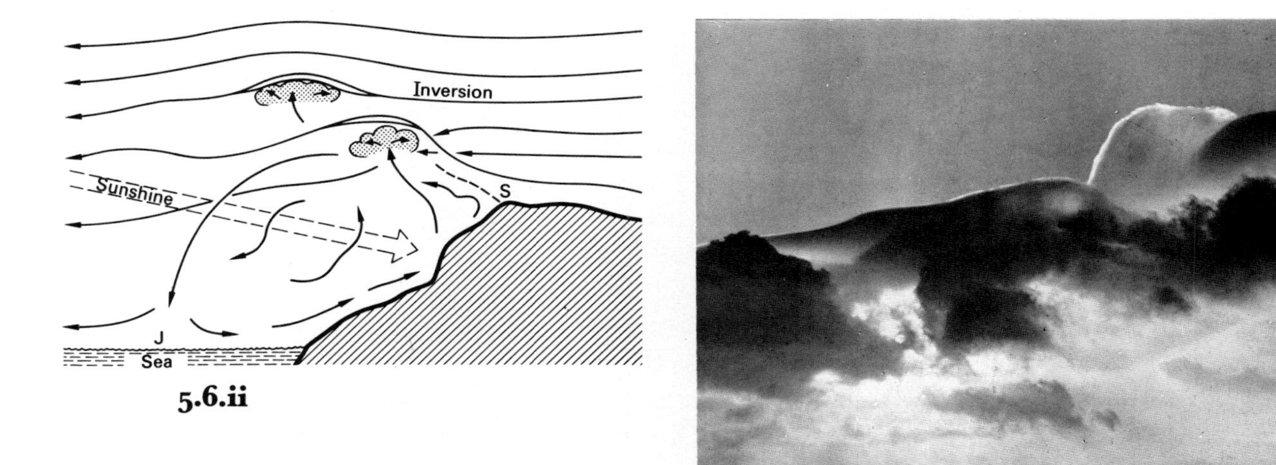


5.6.ii

5.6.4

**5.6.5** follows 5.6.4 by two minutes. On the left can be seen smoke rising from heather burning on the lee slope (see 5.3.3). This type of flow could well be taking place in 5.6.3 but with invisible thermals. Separation from the top of a lee slope is always enhanced by anabatic flow, and this produces unsteady waves in the stable air above.

**5.6.7** In (a) two thermals have just produced small cumulus clouds near to the camera. Further away there is a wave cloud which has been thickened in the centre of the picture by an anvil of a large thermal which has spread out at the inversion. The base of this anvil can be recognised by the light and dark shades showing wrinkles on its under side. Beyond are more thermals.

A minute later the anvil cloud is seen in (b) to have moved downwind from the stationary wave cloud. Because of the added moisture there the anvil cloud does not evaporate in the wave trough which does evaporate the lee edge of the wave cloud.

**5.6.6** In the morning to the west of Mt. Blanc where there are many deep steep-sided valleys, when the slopes are warmed in the sunshine the motion is very unsteady as anabatic flow begins and separation occurs near mountain ridge crests. These three waves moved along, almost with the speed of the wind, riding through the crest of a wave of much greater wavelength. The clouds are pileus-like but with many holes and ragged fragments, as if a dry thermal had penetrated the moister layer of the pileus. These clouds are most common in the morning when a smooth wave motion changes rapidly just before cumulus appear. The sky was darkened by the use of a polarising screen.

Meanwhile the thermals nearer the camera have grown upwards, and have also moved downwind towards the wave cloud, and the cloud shadows have moved. The wind is from the west, the view eastwards in mid morning in early September in the Pennines of Northern England.

## 5.7 Separation

**5.7.1** This famous photograph, originally in black and white, has been coloured by an artist using colour pictures of the region taken at other times when the phenomenon was less striking (see figure 5.7.i on next page).

The Owens Valley (at 4,000 ft.) is dry in March in the lee of the Sierra Nevada of which Mt. Whitney is the highest peak (14,495 ft.), seen on the right. After the descent of around 9,000 ft. the strong wind blows up dust from the valley floor and this is carried up to around 18,000 ft. where there is a cloud in the first lee wave above the rotor. The rotor is filled with dusty air circulating in it. High above the rotor is a thin wave cloud which appears to be glaciated. The Inyo Mountains are under the downwind side of the rotor.

This occasion was remarkable for the strength of the upcurrent in front of the lee wave. Robert Symons soared a fighter plane, with engines stopped, from 13,000 to 31,000 ft., and this picture was taken during this ascent. The 'dust wall' is exceptionally well developed. At that time (1950) there was no theory of rotors and the mechanism whereby the dust was carried up so steeply was not understood. When the picture was first published the heights of the clouds in the wave were deduced on the assumption that the picture was taken at the top of the climb (31,000 ft.) but this now appears to have been an error: since there is cloud at around 14,500 ft. over the mountains it is not possible for condensation to be delayed in the lee wave beyond about 18,000 ft. by any likely mechanism. Even to delay it to this altitude it is necessary to suppose that the cloud over the mountains was glaciated and that the air is warmed by about 8°C (it was midday) or very much mixed with drier air above it in crossing about 14 miles of valley floor below the snowline: the only other possibility is that the air containing the cloud over the mountain was stagnant, being blocked by the mountains, and that it is the air from above that cloud that is descending the lee slope, with only some of the cloud-containing air flowing through passes between the peaks.

The flow is said to separate from the surface where it converges from both sides and rises off the ground (or water). The flow is said to rejoin the surface where the air descends to the ground and spreads out sideways. In each case the wind is zero at

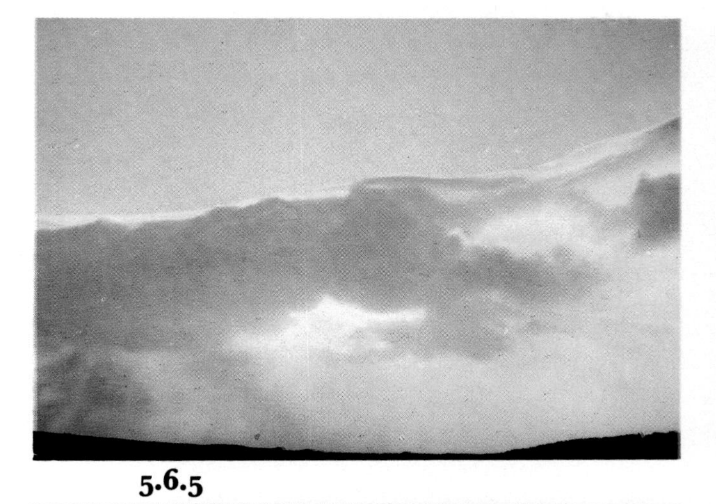

5.6.5

5.6.6

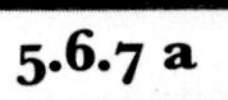

5.6.7 a

5.6.7 b

5.7.1

separation (S) or joining (J). There is an important difference: at a line of separation the flows from the two sides may have unequal components of velocity along the line, in which case the convergence may intensify the shear and produce intense vortices; or the two streams may have different temperature, or moisture or dust contents, so that the streamline from a separation point may be a sharply defined boundary which is clearly visible. At points of rejoining of the flow there is not usually any such phenomenon, and when the flow descends to the ground the most striking effects are produced at the advancing front of the cold or descending air (see 3.3).

**5.7.i** Separation in lee waves occurs when their amplitude is large. Under each wave a rotor is produced with separation at the front and joining at the rear: in between there is a closed circulation.

Other aspects of lee waves are also illustrated in this diagram.

1. When stable air descends from high ground a strong wind blows down the lee slope, sometimes called a Helm wind or Foehn wind (H).
2. The wind at the ground is reversed under a rotor but is usually not strong. In between rotors the surface wind may be very strong (G) with a rather sudden falling off towards the separation point (S).
3. In many cases the streamlines at higher levels are carried higher in the first lee wave than over the mountain so that a cloud may appear there (L). In this case we have illustrated also the possibility that some of the particles freeze and form a trail of orographic cirrus which will extend indefinitely downwind as long as it remains above the ice evaporation level of that air.
4. A second lee wave with a rotor of equal size is very rare: there is often a second mountain range which changes the flow, or the turbulence created in the first wave, when its amplitude is large, is such as to diminish the amplitude of the second.
5. If a cloud is formed at R in the rotor it seems to be tumbling over because of the light, or even reversed, wind at its base. Such clouds are always extremely turbulent because they are formed in air that has a dry adiabatic lapse rate which is produced in the bottom part of the rotor where the layers are the other way up.

**5.7.ii** When the wind blows up a steep hill face separation often occurs if there is a salient edge S at the top. The size of the eddy formed depends on the slope of the ground, among other things, and the point of rejoining, J, may move about erratically.

There are many gliding sites on hills where the upslope wind is used for soaring. Often, to land on the hill top, the glider may approach through a downdraft above J. These are known as 'clutching hands' because of the sensation to the pilot. Such downdrafts are not always associated with the descent of air to a point of reattachment of the flow, but may be vortices developing on the streamline rising from S.

**5.7.iii** When the top of a mountain ridge is sharp separation almost invariably occurs at the salient edge. The streamline from the point of separation is often marked by the top of a cloud (5.7.3 and 5) formed in the air ascending the lee slope.

**5.7.2** The wall beyond the wind socks is at the top of a steep wind-facing slope. The wind socks are at 2 metres and 4 metres and show the size of the eddy in this case. The flow is indicated in figure 5.7.ii.

**5.7.iv** When the mountain is rounded the point of separation moves about (5.7.4); it usually reaches nearest to the mountain top when anabatic flow develops by day (a).

If a sea of cloud exists below the mountain peaks with a very stable layer at its top the flow may be like (b) (see 5.7.4 and 7). Because of the ascending motion generated (sometimes anabatically) close to the mountain side the cloud top can be seen to be shorn off where the wind blowing down the slope leaves the surface (S). On the other hand eddies may be formed by separation at a higher point if the slope is steep and cloud fragments are seen to be carried up out of the layer (5.7.6) from time to time if the cloud top is not very stable (c).

**5.7.v** When the wind is strong and a velocity discontinuity is generated at the line of separation (SS) on the lee side of a hill vortices are sometimes produced which form visible whirls of dust or snow (5.7.8). Lumps of turf have been seen whisked up into the air to a height of 10 or 15 metres by these whirls. If they occur on a water surface, which is rather rare, a 'water devil' can be formed.

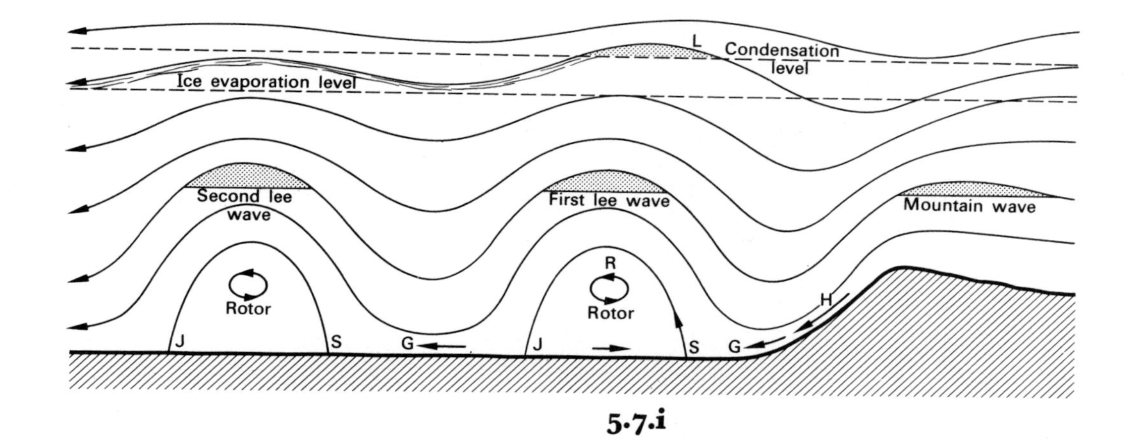


5·7·i

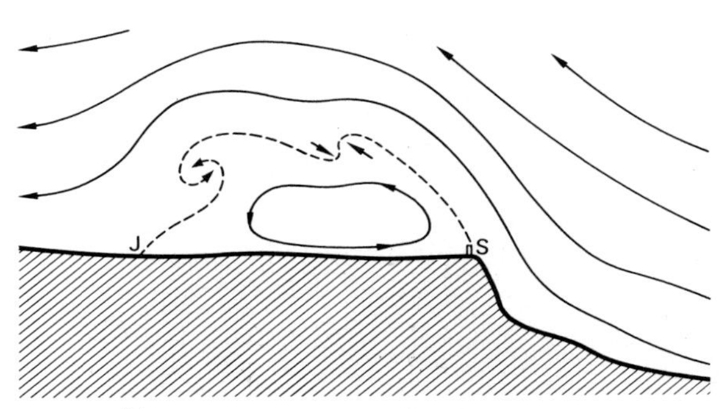


5·7·ii

5·7·2

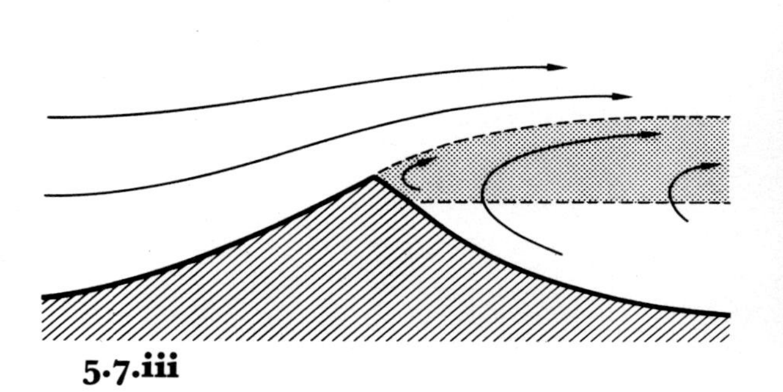

5·7·iii

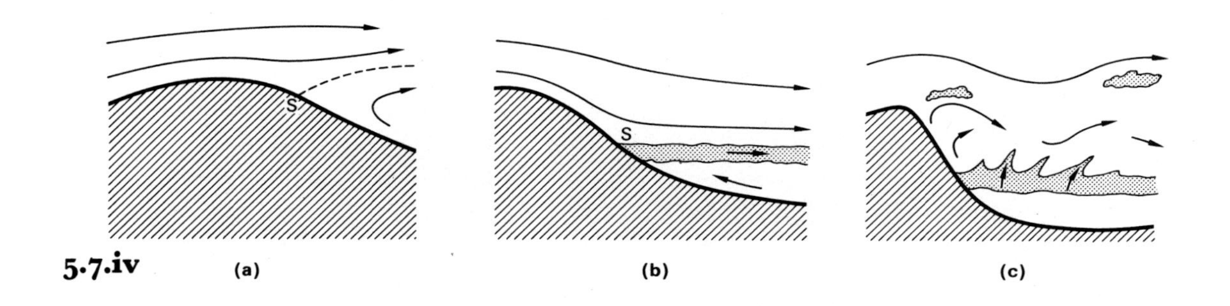


5·7·iv (a) (b) (c)

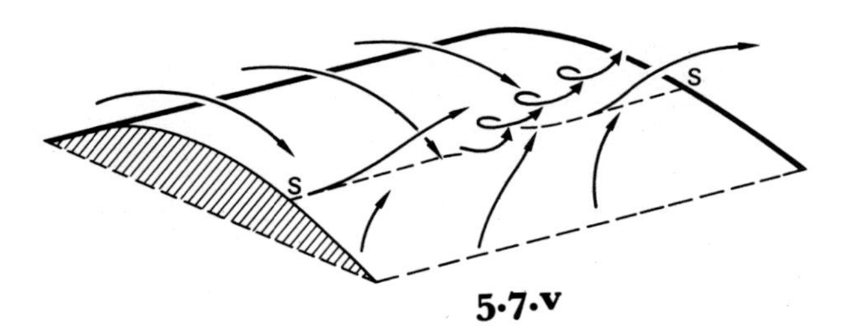


5·7·v

**5.7.3** An observatory and television station occupy the crest of the Pic du Midi in the Western Pyrenees. On this occasion a wind from the south separated at the crest and the air ascending the north slope contained cloud (see also 5.9.2).

**5.7.4** The point of separation on this rounded mountain top (Black Mountains, South Wales) moved around so that the smoke from the generator sometimes travelled down and sometimes up the slope. It could be followed right across the valley at around mountain top level, showing that the whole valley was filled with an eddy. The clouds above showed no waves, and the gaps in them moved along with the wind.

The wind up the mountain side increased upwards from the surface in the first few metres, and consequently it was easy to hear the stream flowing in the valley bottom: the refraction of the sound due to the wind shear made an acoustic duct along the ground.

**5.7.5** It is common for the air masses on opposite sides of a mountain to have different condensation levels (e.g. 5.7.3). Here we see cumulus on the west side of Cader Idris (North Wales) formed below the mountain top while the wind flows from the ESE and carries the cumulus tops away from the observer (looking NW).

**5.7.6** The eddies shed from a salient edge at a mountain top caused erratic motion in the top of a sea of cloud on the lee side. 'Fragments' of cloud rather similar to those in 5.6.1 are seen on the top. This picture was taken about an hour before 5.7.3 and 5.9.2, before the anabatic flow up the lee side had carried air, formerly in the sea of cloud, to the mountain top [see figure 5.7.iv (c)].

**5.7.7** Eddies below the point of separation are sometimes visible in snow or other material blown up from the surface. This is a winter picture in the Rockies of Colorado. There is thin cloud forming over the ridge, close to the ground which is at about 3,000 metres.

**5.7.8** The stretching of the shearing layer where separation occurs on the lee of a mountain can intensify the rotation enough to produce a 'snow devil' (see fig. 5.7.v). The motion differs from that of dust devil (14.1.1) in that the air is not unstable at the surface and the upward motion is not due to thermal convection.

**5.7.9** is an example of dust being raised as the variable point of separation in front of a rotor in the Owens Valley near Bishop (see 5.7.1) moves across loose alumina. In this case, where there is intense warming of the surface, the convergence towards the separation point could concentrate the convection there and intensify the updraft locally.

**5.7.10** Cloud appears to stream almost continuously for large distances downwind from the large peaks of the Himalayas. This is Kanchenjunga (8,586 metres) seen from about 50 miles away to the south in a west wind. The cloud through which the peak protrudes is layer cloud at this late hour (sunset, February) yet the plume reaches at least 1,800 metres above the peak. It is probable that by contact with the frozen surface, from which particles are carried by the wind, the cloud is completely glaciated and that this is responsible for its penetration to such heights.

5·7·3

5·7·4

5·7·5

5·7·6

5·7·7

5·7·8

5·7·9

5·7·10

**5.7.11** At the time of the sunset picture 5.7.10 it was possible to see from Tiger Hill (3,000 metres) the operation of the katabatic wind in preventing separation from mountain surfaces. Anabatic flow, especially when the sun shines on a hillside (5.7.4 and 5) always induces separation at the top: katabatic flow occurs when the ground cools at dusk and the stable layer flows down the lee slope. Provided that there is not a salient edge to induce separation this causes the flow above also to follow the ground contours. The cooling produced this thin layer of cloud on the rounded hilltop and it could be seen, in the moderate west wind, to be moving rapidly over the top and down the lee slope, where it evaporated as a result of descent (see 9.1.4).

This mechanism has some remarkable effects. When the separation is inhibited near a ridge crest in the evening the katabatic flow induces a helm wind to blow, and this helm wind has often been thought to be the katabatic wind itself. Consequently much too great intensities have been ascribed to genuine katabatic winds caused by surface cooling. Helm winds have been known to cause blizzards in valleys in Greenland and Antarctica and their onset has been sudden when separation rather suddenly ceases.

The cessation of separation in the evening and its onset in the morning is the cause of the appearance and disappearance of wave clouds thousands of metres above the ground at those times of day.

**5.7.12** When a mountain protrudes through a layer of cloud with a very stable top the flow may separate in its horizontal motion leaving a region of stagnant, subsided, stirred, or heated air in the lee which has no cloud in it. In this case the clearance is in the lee of Fuerteventura (Canary Islands). The same very thin cloud layer is seen in 1.6.8 where it is cleared over the African mainland by sunshine. A similar cloud also appears in 7.3.1. The remarkable feature in this case is the very sharp eastern edge of the gap. Often the vortices produced on the separated streamline extend for hundreds of miles downwind (see 14.2.2).

**5.7.13** This lane in a cloud layer extended from the SE corner of Ireland (Carnsore Pt.) across St. George's Channel towards N. Wales in a wind from the South West, approximately along the lane. The distant clouds are over Snowdon. The layer of cloud is waved where it is in the lee of Ireland; to the right of the gap it is not. The motion causing this gap is not known for certain. The distant curvature of the gap may represent a changing wind moving the gap sideways, rather than a curved flow. If a westerly component of wind was increasing the waves might be produced by the mountains of Ireland.

**5.7.14** The sunshine on a hillside causes the air close to the hill to become warmer than the air at the same altitude over the low ground. Consequently an upslope (anabatic) wind occurs.

Cloud at a lower level than the main condensation level is produced by evaporation of moisture from the ground in the sunshine. The cumulus so formed evaporates rapidly as it rises and grows in the drier air above; in the meantime it makes the motion visible and indicates the region of air heated by the warm ground.

On reaching the top of the slope the thermals rise vertically.

**5.7.vi** The anabatic flow up a warmed hillside is confined to the shallow layer close to the slope. The air over the low ground must be stably stratified (the continuous lines indicate constant potential temperature). The flow continues to ascend above the hill tops first because of condensation (see 5.7.14), and later when the whole lower air mass becomes warmer. (Trapped anabatic winds are shown in 9.2.i.) The outflow and downflow over the low ground is an order of magnitude smaller than the upslope flow. By contrast katabatic flow is much shallower and never extends above the hill tops.

5·7·11

5·7·12

5·7·13

5·7·14

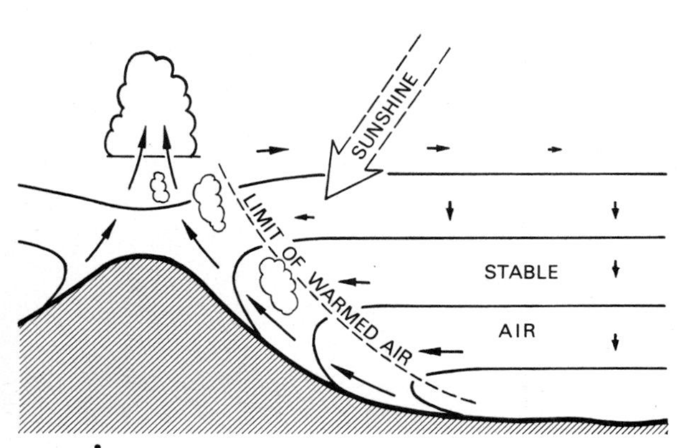



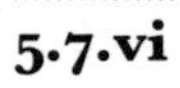

5·7·vi

## 5.8 Wave Cloud Recognition

**5.8.1** When a warm front approaches the thermal wind increases and so the wind becomes stronger at high altitudes. This together with the increase in stable stratification due to the arrival of warmer air aloft, and some subsidence of the cold air beneath (which does not invariably occur), produces conditions favourable for an increase in wave amplitude. The picture shows the first wave clouds appearing in middle levels as the cirrus advances. (See also 5.1.5.)

**5.8.2** This photograph was taken vertically upwards with the sun above the cloud. The left edge is fairly smooth and is where the air enters the wave cloud. Because the condensation in this case makes the air unstable a cellular convection occurs and the humidity distribution becomes lumpy. It can be seen from this that the wind is from left to right. (Compare 6.3.5.)

**5.8.3** At cold fronts the thermal wind and temperature structures are often favourable for wave formation, especially after the rain belt has passed. In this case waves are formed over Snowdonia as the front passes away to the NW. The jet stream is parallel to the main cloud edge.

**5.8.4** These wave clouds over Norway were measured to be at about 28km, in the stratosphere. Although of the Mother of Pearl type (see 5.4.5) there is no iridescence recorded in this photograph. But the contrast in texture between the smooth bright water cloud and the tenuous trail of ice crystal cloud extending for a great distance downwind is very well illustrated. Below are tropospheric clouds in darkness after sunset. (Compare 5.4.1.)

**5.8.5** In the evening when katabatic winds begin to make the wind follow the contours of the ground more than during the day wave clouds are often seen for a short time before sunset. They are frequently variable in shape and wavelength and unsteady in position. Here there is a lower layer of dark waves with an upper layer with cellular structure and longer wavelength above. The cells appear on one side, and evaporate on the other as the air descends into a wave hole.

**5.8.6** This shows the last smooth waves to occur in the morning as convection developed and formed cumulus at a lower level. The small wave on the right has finger-like corrugations (see 6.1.2) on its left end. These lie along the direction of the wind. This type of cumulus is often called stratocumulus, but is not really formed by the spreading of cumulus. It tends to remain predominantly located over the hills, like wave clouds.

## 5.9 Billows in Waves

The mechanisms whereby billows are formed are described in the next chapter. These two pictures illustrate one mechanism in action in waves.

**5.9.1** As the ground cooled in late afternoon when the shadow of the cloud in the top of the picture extended across it a wave cloud not disrupted by convection appeared at about 500 metres over the mountain in an onshore wind. For several minutes billows appeared on its upper surface, having the appearance of breaking waves which collapsed into the cloud on the right. The spacing is 150–200 metres and each billow was observable for about 2½ minutes.

**5.9.2** The upper surface of the cloud in 5.7.3 had billows forming on it visible for several minutes before it was evaporated by the sunshine. This close-up of two vortices shows small billows forming 'on the back' of the vortex on the right. The cloud top represents a density discontinuity which becomes unstable when tilted and is rolled up into vortices. The tilting in the right vortex induces the smaller ones.

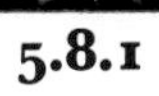
5.8.1

5.8.2

5.8.3

5.8.4

5.8.5

5.8.6

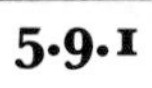
5.9.1

5.9.2

# 6 BILLOWS

## 6.1 Billows in Waves

One of the mechanisms which produce billows is the intensification of shear on a tilted discontinuity of density. It has been well demonstrated by Dr. S. A. Thorpe whose experiments are illustrated here.

Cloud tops are often much colder than the air above because of the loss of heat from them by radiation and when these are tilted in mountain waves enough shear is often generated to cause the layer to break into a row of vortices, and the cloud top has the appearance of breaking waves. (See 5.8.1, 2).

The second billow mechanism is shown in figure 6.1.iv.

**6.1.1 a, b, c** These three experiments show features of the billow formation mechanism. In (a) there is dye in a layer of intermediate density. Between the top and bottom layers, which are clear, there is a 7 per cent density difference. The total tank depth is 10 cm. and it was at rest horizontally. It was then tilted to the inclination shown here so that the lighter fluid ran upwards and the lower fluid downwards. The dyed layer thus became a vortex sheet. Where it is displaced upwards or downwards it is moved, by the velocity field, to the right or left respectively so that the layer is thickened where it slopes against the tank tilt and is thinned out where the slope is exaggerated (see figure 6.1.i). The concentrations of the fluid of the vortex layer are also concentrations of vorticity and they soon cause the surrounding fluid to rotate around them. Later stages in the development are shown in (b) and (c) in which there are only two layers, the lower one being dyed.

In (b) the total depth is only 3 cm. and the billows occupy a large fraction of the whole tank depth. In (c) the total depth is 10 cm., yet the billows are only slightly larger than in (b), which is less reduced in the photograph. The spacing of the billows in (c) is mainly determined by the thickness of the shear layer generated by the tilting.

There are three possible fates for billows of this kind:

(i) after rolling over, as in (b) the fluid becomes mixed and flattens out as a new transition layer of greater thickness and therefore of smaller gradient. This layer would produce billows of greater length if subsequently tilted sufficiently, or

(ii) the billows will persist in a pattern of motion known as cat's eyes (see figure 6.1.ii), with a wavy motion above and below, or

(iii) by the time the billows have reached only a small amplitude the air has moved to a different part of the mountain wave which produced the original tilt of the stable layer. Consequently the vorticity becomes reduced and the layer no longer unstable. This is most likely to happen where the billow wavelength is large (e.g. in excess of 200 m.) for then the growth rate at the unstable stage is small. In such a case the billows will not grow beyond a certain point, and the cloud will show transverse corrugations but no rolling over.

Occasionally the waves may become cusped at this stage with some cloud carried off as a thin wavy layer from the top of the cusps. [See 6.2.5 and 6.1.i(5)].

**6.1.i** The growth of unstable waves in a vortex layer may be thought of as follows:

1. When the layer is undisturbed the vorticity transports the upper and lower layers in opposite directions with uniform speeds.

2. The crests and troughs of any transverse corrugations are transported in the direction of the layer into which they penetrate.

3. The vorticity is accumulated by the thickening of the layer at alternate nodes of the original wave form. The layer is thinned at the other nodes.

4. The layer becomes rolled up into transverse billows. If the bottom fluid only is visible the appearance is rather like 'breaking' waves (but the mechanism is actually quite different from the breaking of waves on a sloping beach).

5. If the instability is checked by a reduction in the velocity difference between the fluids, the vortex layer does not roll up, but may have cusped waves on it from which thin filaments of the lower layer are carried away embedded in the upper layer.

These waves are not symmetrical and no material of the upper layer penetrates into the lower. For this to happen the upper fluid must usually be mixed up by turbulence while the flow in the lower fluid is laminar.

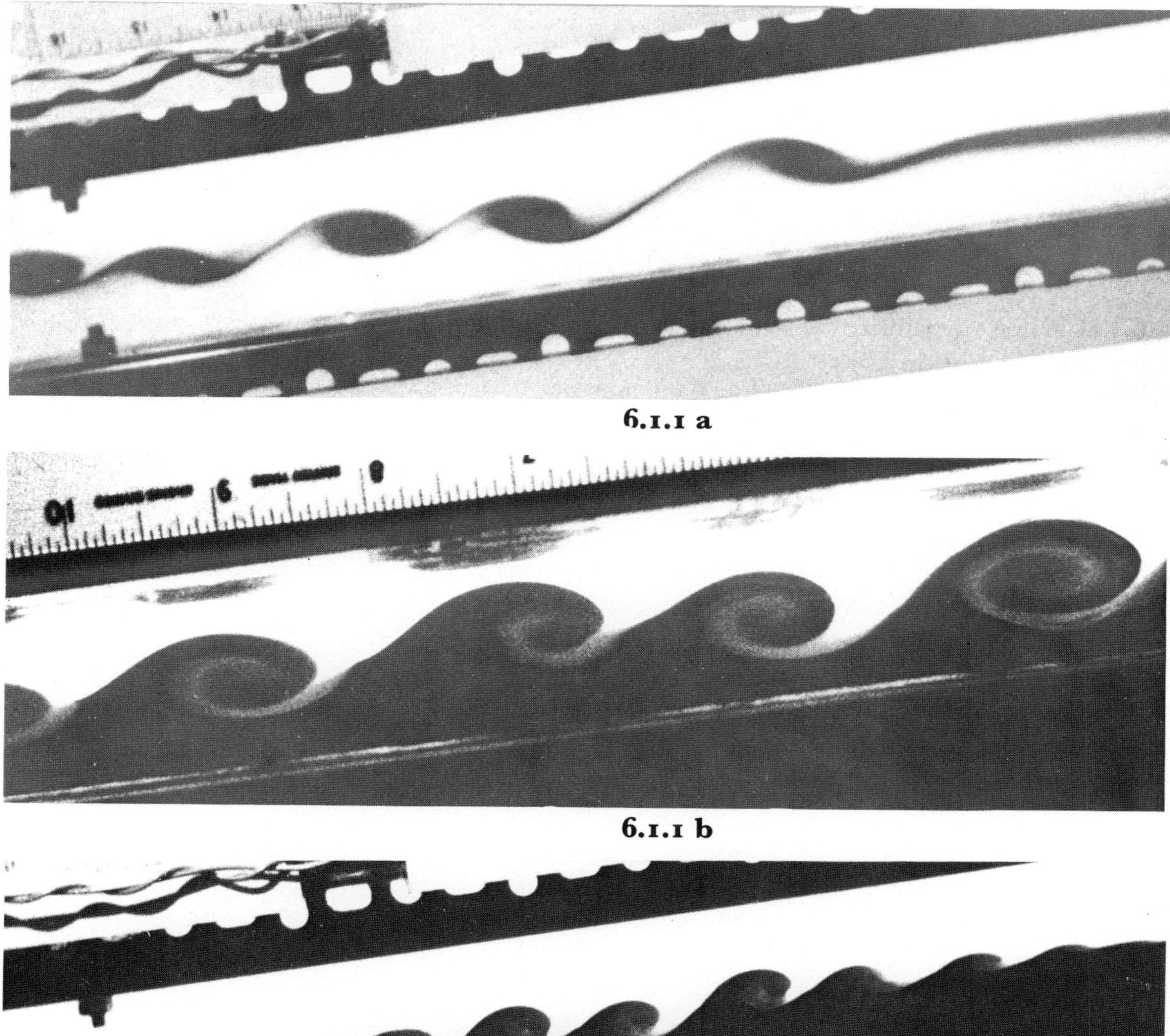

6.1.1 a

6.1.1 b

6.1.1 c

**6.1.1d** The waves shown here move to the left with the upper fluid and thin filaments of the lower fluid, of which the top part is dyed, are carried up rather like the breaking of large amplitude waves in which the crests travel faster than the troughs.

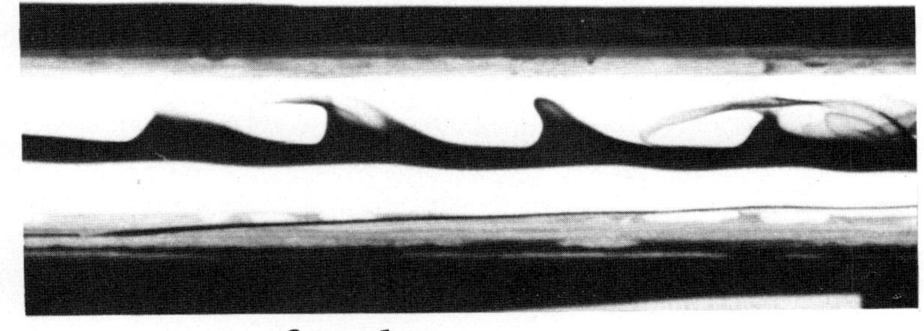

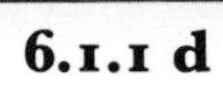

6.1.1 d

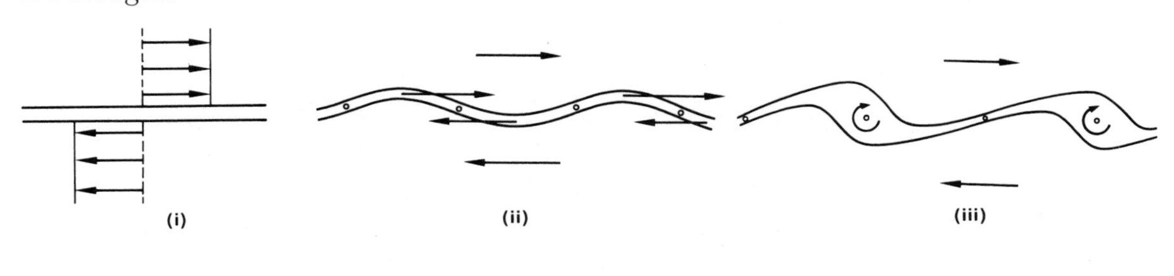


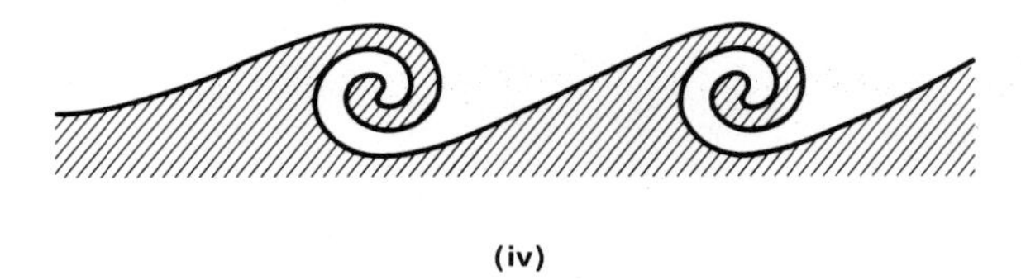


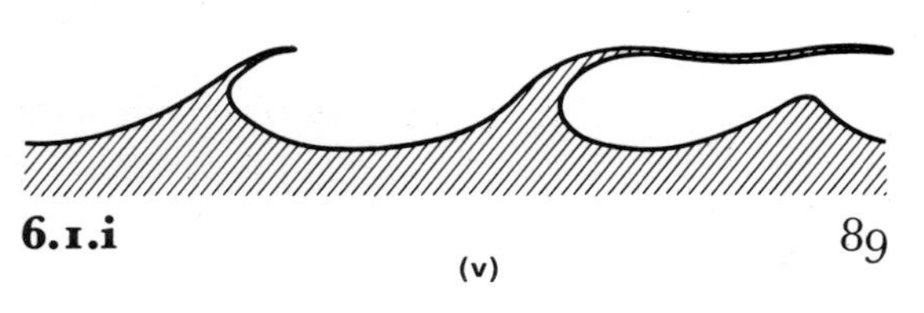


6.1.i

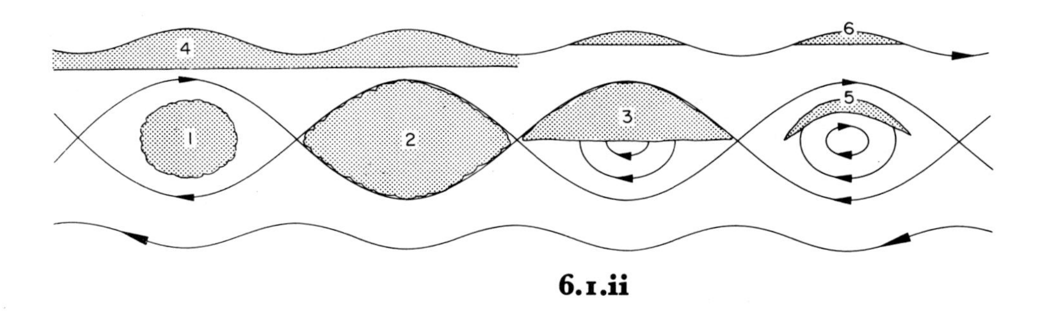


**6.1.ii**

**6.1.ii** When the rolling up is complete the motion acquires a cat's eyes pattern, which has a wavy motion above and below it. The spacing of the eyes is about 2·7 times the depth of the layer which is used to fill the closed circulations. What sort of cloud will be produced depends on where in the pattern the cloud occurs. The diagram shows six possible shapes.

1. rolls in the centre of the eyes, with clear spaces in between,
2. rolls occupying the whole of the closed circulations, with almost no gaps in between,
3. rolls in the upper parts of the circulation only i.e. above a condensation level,
4. a layer with a corrugated top, the base being at the condensation level,
5. 'eyebrow' clouds in the upper parts of the circulations,
6. eyebrow clouds in the wave motion above, with arched or flat bases.

Other forms are also possible (see 6.5).

The growth of the waves may be significantly reduced, when the wavelength is large, by the stable stratification in the air above and below, for growth requires energy for the vertical displacement of the air, which may exceed that available in the instability. In stages 2 and 3 of 6.1.i the wavy motion in the air above and below has the same phase as in the vortex layer. In the cat's eyes pattern the troughs in the upper layer are above the crests of the waves in the lower layer. There is at present no complete mathematical theory for this change, nor for the instability when the vortex layer is not very thin. It is known that the thicker the vortex layer the greater must be the velocity difference between the upper and lower layers before instability occurs, and the longer will be the wavelength.

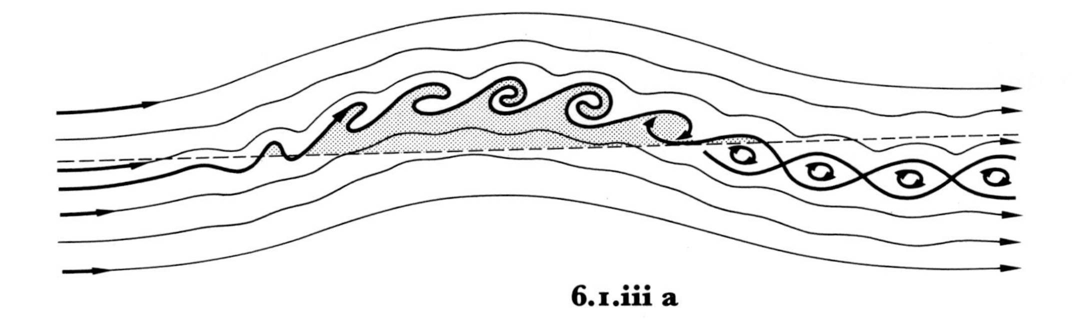

**6.1.iii a**

**6.1.iii (a)** Since vorticity is increased in the forward direction when the flow is tilted upwards it is a maximum in wave crests, which is where billows are most likely to be well developed. If they 'break' they will produce a cat's eyes pattern of motion which, being stable, may persist for some distance downstream.

**6.1.iii b**

**(b)** the reverse of this billow motion must necessarily occur in a trough, but it is unlikely to be visible in clouds because they will be evaporated there.

**6.1.iv** Where the motion is sufficiently curved and the velocity gradients large, centrifugal forces can play a part similar to gravitational forces. Where a strong shear layer enters wave motion the faster moving fluid tends to move to the outside of any curve in the flow. Thus if the faster fluid is on top it will tend to move under the slower moving fluid beneath it in the trough of a wave. This kind of instability takes the form of longitudinal corrugations, and it can only occur when the curvature and shear together are enough to overcome the static stability. The figure shows corrugations growing as the air proceeds through a wave trough.

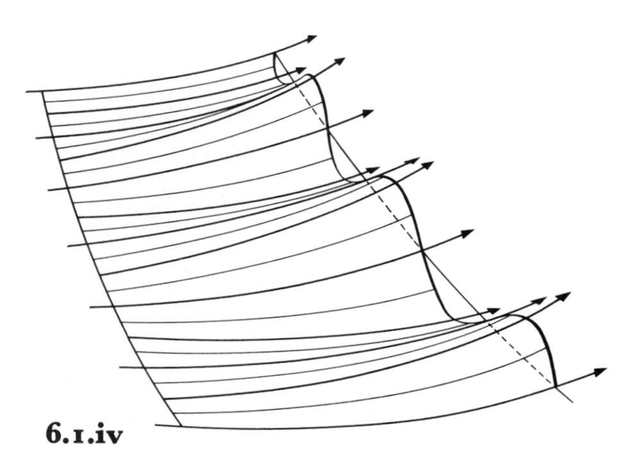

6.1.iv

As soon as the air enters the next wave crest a formerly unstable layer becomes very stable, so that the corrugations spread across the cloud as stable waves, ready to grow again in the next trough. In this way they can persist for some distance in an area of waves.

6.1.2

**6.1.2** This view is from about 3,000 metres above a wave hole near Camphill (near Sheffield) seen from a glider. The wind is from the left and the corrugations can be seen on the cloud surface, extending like fingers into the wave hole.

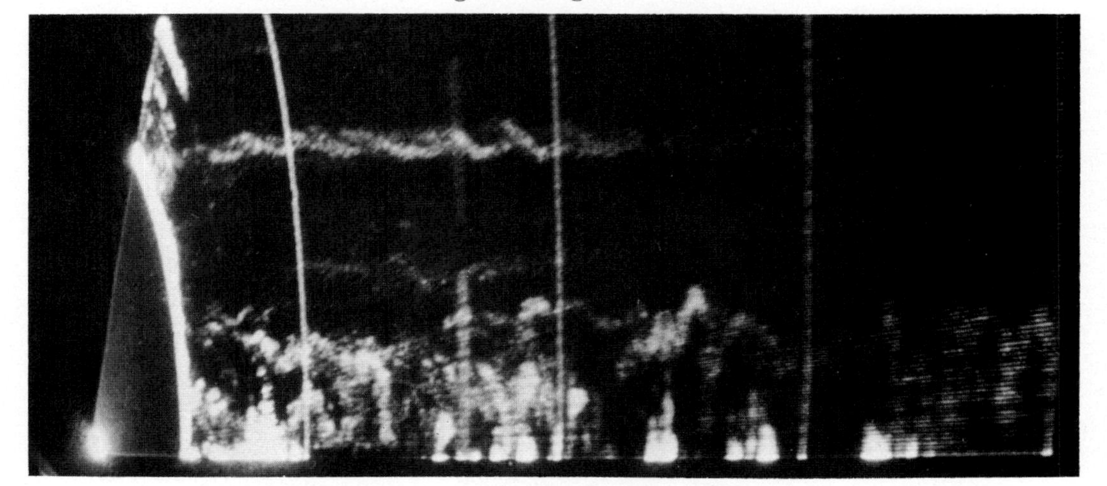

6.1.3

**6.1.3** Billows may occur on stable layers which are invisible but which return a radar echo because of a large gradient of humidity in the layer. In this case a billowed layer can be seen at about 3·5 km. Below about 1·8 km. are thermals rising from the warm ground.

The vertical white lines are at a range interval of 5 nautical miles, and the full depth of the picture is about 4·8 km.

**6.1.4** There appear to be two layers of billows in this wave, one is probably in or near the centre of the vortices while the others are in the wave motion above. The wind is at right angles to the billows from the right. The arched clouds show longitudinal corrugations.

**6.1.5** The wind is from the right through this wave cloud and in such cases the billows can be seen evaporating along the lee edge (left and top in picture). The upwind side (low right) is mostly free from billows.

**6.1.6** In this wave the wind is from low left along the streaks which extend through the cloud. These may be longitudinal corrugations of the type shown in figure 6.1.iv. There are scarcely developed billows at right angles to this direction.

**6.1.7** This is part of a large wave cloud seen looking westwards in the early morning. Through the gaps between the billows we see the brightly illuminated tops of the rolls of cloud.

At the bottom of the picture we see a contrail and the shadow of a nearer part of it is cast on the cloud in the low right hand corner.

**6.1.8** Billows move into wave clouds and out on the far side with the wind. Here we see some curved billows, a rather rare phenomenon. It is probable that the humidity distribution existed already in the air from a previous billow motion, and that it had been distorted in the meantime to produce the arcs.

**6.1.9** When the waves are of small amplitude and formed over a complicated hill pattern, shear may be generated in almost any direction. If this is added to shear already existing, the instability can appear with various orientations. It is unusual for billows to appear on the base of a cloud layer because this is not a stable layer unless the layer is formed by the anvil mechanism or a similar upsliding motion of a frontal surface. More usually the appearance of cloud base billows is given by billows in a thin cloud layer lying close below a dense cloud layer. These billows were seen in clouds beginning to thicken ahead of a warm front soon after 13.3.1 over the small hills of central Lincolnshire.

**6.1.10** Large cumulus were growing rapidly to form anvils over S.W. Scotland. One glaciated anvil is seen on the extreme right. The cumulus seen here is building up again over Cumberland and in the distance over the mountains the anvils which had disappeared over the water reform as a wave cloud over the mountains SW of Carlisle. The view is from Gretna Green. (See 5.6.7).

Billows of rather long wavelength are formed (probably 1–2 km.) which have the same appearance as short wavelength lee waves, but which in fact travel with the wind through the main cloud mass.

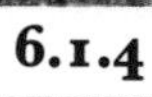

**6.1.4**

**6.1.5**

**6.1.6**

**6.1.7**

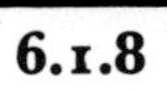

**6.1.8**

**6.1.9**

**6.1.10**

## 6.2 Billows in Layers

We have seen that a slight waviness in the flow can produce billows in layers. They can also be created when radiation from the top surface of a layer causes static instability in the layer. The convection takes the form of transverse rolls if there already exists enough shear in the layer to prevent the formation of oppositely rotating longitudinal streets. Streets are very rare in layers of cloud because the convection is more nearly a once-for-all overturning which occurs at the moment the cloud becomes unstable, and is not a continuous overturning as in a layer being progressively overturned in sunshine over warm ground.

**6.2.1** The entrance to the stronger part of a jet stream is accompanied by the ascent of warm air: this is necessary to provide the kinetic energy for the increased wind. The precise place at which cloud first forms may be determined by a small range of hills for an hour or so. Here we see the first point of condensation of the clouds as smooth wave clouds which are full of billows. Two quite different wavelengths can be seen on the left. This cloud showed bright iridescence in the unbillowed parts, but this is difficult to show in a cloud of generally very bright illumination. The edge of this extensive sheet has all the characteristics of a wave cloud.

**6.2.2** This layer lay in very light wind midway between England and France. It shows a very characteristic billow formation on the near edge which is not obviously explained. The convection in the layer is due mainly to heat loss from the top and the billow size is determined by the layer thickness and is oriented across the shear direction.

**6.2.3** Thin layers of cloud have a much darker appearance than growing cumulus because much of the incident sunlight is transmitted. This is because the clouds are thin and also because they contain no very small droplets because at the small rate of condensation occurring, all the excess moisture is taken up by the larger ones. In cumulus on the other hand all the nuclei form drops and the cloud is very opaque with the sunshine brightly reflected.

The delicate cumulus puffs (over Poitiers) are contrasted with the thin lacework of the billowed layer. This example gives a perception in stereo that cannot even be seen from the aircraft.

**6.2.4** These billows over the wastes of northern Saskatchewan are in a very thin layer about 3,000 metres above the cumulus. In the low right centre they are thickened in front of the rather whiter patch of stratocumulus and in single view appear almost to be part of the stratocumulus layer. High above, almost level with the aircraft is a dark, very thin layer with some darker patches at the tropopause just below the aircraft. The shadow of thicker parts of it can be seen on the layer below. The view is northwards in the afternoon.

**6.2.5** The remarkable feature of these billows, which occurred at the tropopause at about 10,000 metres (viewed from about 12,000 metres) in wave motion in a sub-arctic jet stream over central Saskatchewan in October, is that the waves, instead of appearing to break by rolling cloud around the vortices, has carried the raised cloud over the tops of the vortices. The fifth from the right has an eyebrow cloud above that carried up from the main layer, and it and the billows beyond it have an appearance similar to the unsteady wave clouds in 5.6.1, which is quite common at the moment billows are first forming. But evidently these billows do not fully overturn: they appear to be more like the cusped waves shown diagrammatically in 6.1.i 5 (page 89). The 'shadows' in the corrugations nearer than the main billows are intensified because the cloud is absent in the billow troughs. There is a wave hole near the bottom of the picture.

**6.2.1**

**6.2.2**

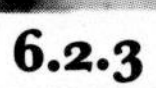

**6.2.3**

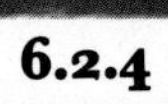

**6.2.4**

**6.2.5**

## 6.3 Complex Structure

A simple uniform billow pattern is most usual when their development is rapid. Some complexities or rarities of structure are illustrated here.

**6.3.1** The rather disorganised pattern of billows in this layer is characteristic of a rather small shear in an unwaved layer. It was about 500 miles west of Ireland and is seen from 7,500 metres. Patches of billows of longer and shorter wavelength can be found with a variety of orientations, none of the motions producing these patterns is as intense as in billows with a regular formation.

**6.3.2** These billows were observed over a mountain in Westmorland and are produced by wave motion. There is some doubt about the position of the thin strips of cloud relative to the broader billow clouds; but they are probably in a very thin layer of cloud above the main billows which is only visible when viewed tangentially (6 in 6.1.ii).

**6.3.3** The double structure in these is probably illusory. The dark strip in the middle of the first and fourth billows (from the right) is the thin edge of the cloud viewed tangentially. The sun is rather low in mid winter so that these edges may be in the shadow of the next billow, or they may be dark like the billows in 6.2.3.

The tops of some (notably the fourth and fifth) are curling over at the moment of the picture.

**6.3.4** The spacing of the billows in this wave cloud near Mt. Blanc varies considerably across the cloud. There are small holes in the cloud, especially in the top part in the centre of the picture. The right side of this part of the wavecloud was evaporating and the holes grew in extent as they moved towards this lee edge.

**6.3.5** An exceptional hole is shown in this wave in the Faroe Islands. The wind is from left to right along the texture of the cloud which may be longitudinal corrugations. The hole, and other instabilities grow towards the right of the picture.

The hole appears like a clear thermal penetrating the cloud and it seems to have caused waves upstream in the cloud as if the thermal were coming from a layer moving more slowly. The same kind of hole is seen in the right of 5.6.6.

The hole is probably penetration of clear air from above which has cooled more than the cloud in rising through the wave. Similar but more extensive holes are appearing in the top left of the picture.

**6.3.6** Just above the skyline and in the top right of the picture are two patches of billows. The lower patch was observed in a streak of cirrus which it deformed into a row of arches. The cat's eyes pattern seems to exist below it. This is a view southwards. Thin dark wave clouds can be seen above the mountain on the left.

**6.3.7** This is the same occasion as 6.3.6 looking towards the WNW after sunset. The sun is shining on to the arched billows on the upper cloud base and casting shadows further up the cloud.

The castellatus is formed over the water in the NE corner of Cardigan Bay (Wales): this is a fairly common occurrence when cumulus dies away in the evening over the surrounding land. The ascending motion which produced the castellatus also produced a thin wave-like cloud above it and it is in the same wave that the billows may have formed.

**6.3.8** These billows are making arches in the base of the cloud layer which is illuminated by the distant bright sky off the left of the picture. Because of the low light source the lowest parts of the cloud appear brightest.

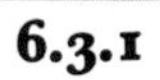
6.3.1

6.3.2

6.3.3

6.3.4

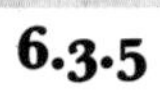
6.3.5

6.3.6

6.3.7

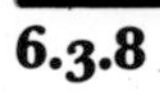
6.3.8

## 6.4 Rolls and Ground Billows

If there is large shear in the morning when convection begins feebly, a transverse motion like billows sometimes appears at the ground instead of streets, which are longitudinal. These are not usually long-lived, and soon disappear as the convection develops.

**6.4.1** These billows seen overhead through a narrow angle lens, close to the sun at about noon, have cloud in the centres of the vortices. Just as a vortex sheet (shear layer) is unstable so also is the system of parallel vortices into which it breaks up: if its regularity is disturbed neighbouring vortices tend to rotate round one another, and this appears to be occurring here. Such motion may be inhibited by the stability of the air above and below or by reversal of the tilt that generated the vorticity which caused the billows. This cloud represents type 1 in 6.1.ii (page 90).

**6.4.2** When the rolls are based on the ground parallel lines of cloud across the wind and shear direction are produced when the circulation reaches above the condensation level, as here. It could be that the billow motion is all taking place well above the ground, but this is unlikely to be the explanation in this case because the spacing is such that the roll motion could probably reach very close to the ground and in a strong wind the shear will certainly originate at the ground. There is no commonly occurring mechanism for generating shear between the ground and cloud base except by means of the ground drag.

**6.4.3** On a morning of slowly developing convection, for a short time the shear which was present before the convection occurred can cause transverse rolls to occur; but these are shortlived, and as soon as the vorticity produced by the convection dominates that initially present, the rolls vanish. Like streets, rolls also need a layer of fairly uniform depth.

**6.4.4** Motion like that of newly forming billows is sometimes seen in the evening or early morning. The best situation is when convection is just beginning in a layer of large shear. This case occurred late in a June evening after a cold front passed over London. The cloud is scud from the wet warm ground and is being distorted by the vortical motion in the air. The dark clouds are like pileus above a vortex.

## 6.5 Cirrus Billows

Frozen clouds sometimes give a fibrous texture to billows. Occasionally the development is so rapid that new supercooled droplets are condensed and the billows look brighter than the surrounding cloud just as an unglaciated wave cloud shines more brightly (see 5.4.1), but this is rare. Usually the ice cloud acts as a tracer of the motion.

**6.5.1** Because of the non-evaporation of ice particles in the billow motion pattern the thicker parts of the cloud are probably mostly in the closed circulations with threads of cloud stretched in between which are just outside the circulations.

**6.5.2** When a layer of ice cloud becomes billowed the cloud is redistributed by the motion. In this case perhaps the most remarkable effect is just to the left of the centre of the picture where the centre of the rolls appears to be empty, surrounded by a tube of cloud.

**6.5.3** Patches of cirrus such as this one are usually notable for the slowness of any changes that occur in them. The occurrence of these billows indicates that the ice cloud was at a very stable layer and therefore more or less at its level of formation: the particles therefore had scarcely fallen at all.

**6.5.4** These billows, in which there is one rather long dominant wavelength and two or three smaller, differently oriented ones, was observed at Font Romeu in the eastern Pyrenees soon after sunrise in August. As the air flowed over the mountains there must have been shearing motions generated in many different directions, and each billow pattern generated left its mark.

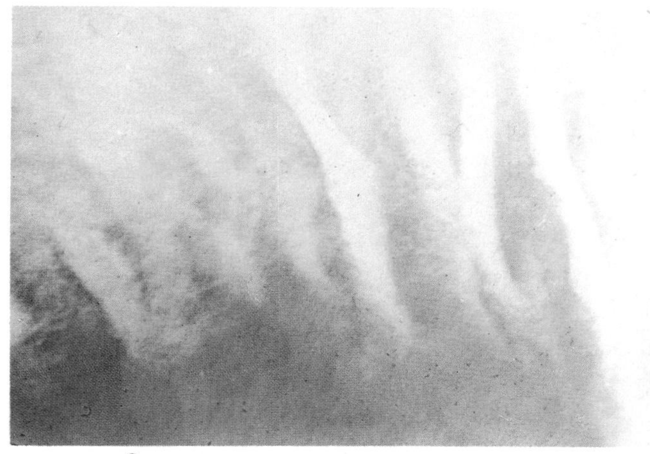
6.4.1

6.4.2

6.4.3

6.4.4

6.5.1

6.5.2

6.5.3

6.5.4

# 7 ALTOCUMULUS

Altocumulus is a layer of cloud composed of cells or lumps. It may be formed in the first place either by widespread ascent at a front, by lifting of the air in a wave, or by cooling of the air at a haze top. Once formed it may sprout upwards as castellatus (see 1.4 and 5.5), in which case the size of the cloudlets depends on the thickness of the layer formed, or it may acquire a billow structure (see 6.1.9, 6.2.3, 6.3.1 in particular), or it may develop very flat cells with a smaller cellular structure superposed, or it may become glaciated partially or completely. Each of the following examples illustrates more than one mechanism so that the classification is not unique.

## 7.1 Upward Growth

**7.1.1** In stereo it is clear that the lumpy cloud on the right is far above the billowed layer. The lumps are scarcely castellatus but their varied size and their appearance as small cumulus which individually evaporate rather quickly show that the instability due to condensation is important in their formation. This was on the edge of a frontal zone over the sea to the S.E. of Newfoundland.

**7.1.2** Thermals from the ground would not produce such small individual cloud elements as these, and the distant edge of the cloud area shows that there is a larger scale motion (probably orographic) producing extensive cloud patches. It is uncertain whether the warmth of the ground plays a part in determining the structure, but it is certainly observed that in hot climates altocumulus tends to disappear by day. This is at Bardai in Tibesti.

## 7.2 Cellular Patterns

The full explanation of cellular patterns is certainly not simple. The model experiments illustrated in this section show an apparent rather than real similarity to cloud forms.

**7.2.2** The edge of the cloud of an advancing warm front was continually extending into the clear air. New cloudlets appeared on the left and amalgamated into larger lumps as the cloud thickened. Evidently the humidity was lumpily distributed because the cells did not display any significant internal motion during the thickening.

**7.2.1** This picture shows two layers, the lower one, seen in the upper part of the picture being thicker and having a larger cell size. The cell size is usually determined by the cloud thickness, but not always (this is obvious from the fact that one cloud layer can have more than one apparent cell size e.g. 7.2.5). Thin layers often have sharply defined edges but their cause is not clear.

**7.2.3** The form of the clouds, particularly in the distance, shows that they are wave clouds, with their overall outline almost stationary. The far end is where the principal growth is occurring but the wind is almost exactly along the lines of small cells so that there is a small component into what is seen here as the left edge of the lines.

There is some similarity to a jet stream entrance, but on a smaller scale (see 6.2.1). Lines of altocumulus whose growth is much less than castellatus, and is not enough to preserve it in a down motion of equal magnitude to the up motion, are often formed in this way. In fact, as in 7.2.2, the lumpy humidity structure seems to exist in some places in the air before the cloud forms and the cloudlets have almost no internal motion.

**7.2.4** The upper cloud is a wave cloud which has two systems of billows feebly apparent in the cells. The wind is towards the observer so that the cloud is evaporating at the top of the picture.

Beneath are cumulus, feebly castellatus, forming in wave crests at the condensation level at which cumulus had its base earlier that afternoon. This level happens also to be a fairly marked haze top level over S.W. London. The altocumulus could well have been at the anvil level of the afternoon's cumulus.

Billows of this type usually only occur in light winds: otherwise in the very slow internal motions of the cells there would not be time during their passage through the cloud for them to be arranged as billows. The lumps are largest where the cloud is thickest.

7.1.1

7.1.2

7.2.1

7.2.2

7.2.3

7.2.4

**7.2.5** The detailed mechanics of these clouds has not been worked out, although they may look simple enough. This is partly because it had been thought for some time that there was a close analogy with cellular motion in layers of viscous fluid. But the typical cell size here, which is about a fifth of the picture width near the bottom, is very much greater than the thickness of the layer cloud. The small lumps within the cells are more nearly the same order of magnitude as the cloud thickness. These clouds are over the Atlantic ocean and shadows of a higher patchy layer can be seen.

The chief cause of heat transfer is the loss of heat by long wave radiation to space from the cloud top. This is sufficient to ensure the continued existence of the cloud, and is greater than the heat gain by radiation from the ocean below. Consequently air below the cloud to some unknown depth must also be cooled by downward convection, and this may be the factor determining the main cell size. (See 1.6.1-2.)

It is evident that there is some small motion in these clouds because it is detectable through the slight vibrations it produces in an aircraft flying through it, but the second reason why so little is known about the internal motion is that it is difficult to photograph: the clouds change so little in passing across the sky. (See Appendix p 169).

**7.2.6** It is very easy to make cellular patterns in a viscous liquid such as white spirit or oil, by placing a shallow metal container over a warm heat source. The motion can be made visible by introducing a small amount of aluminium paint or powder. The flakes of aluminium lie tangentially to the motion because the shear rotates them into this position. The dark lines and spots are where light passes into the fluid between the flakes and therefore represent down and upcurrents respectively.

The motion is very smooth and is limited in magnitude by the viscosity, to which there is no analogous force in the cloud. The loss of heat by radiation at the cloud top is somewhat analogous to the conduction of heat away from the top surface of the fluid, but if there were no mixing with the air above or other mechanisms not present in the model experiment, the cloud would form a very uniform layer without a cellular structure where the air is above its condensation level.

**a.** In this case the oil is of uniform depth of about 3mm. and the cell size is about 6mm., and the heating of the bottom is just sufficient. The pattern may often be very steady.

**b.** When the container is tilted to give a fluid depth varying from about 1mm. to 5mm. the cell size varies according to the depth. The pattern is often very steady.

**c.** When a breath of cool air is blown across the fluid surface a rapid extraction of heat occurs which causes downward convection on a much smaller size scale. The little cells do not extend far beneath the surface and disappear quickly when the increased cooling is stopped. The pattern is rather unsteady.

7.2.5

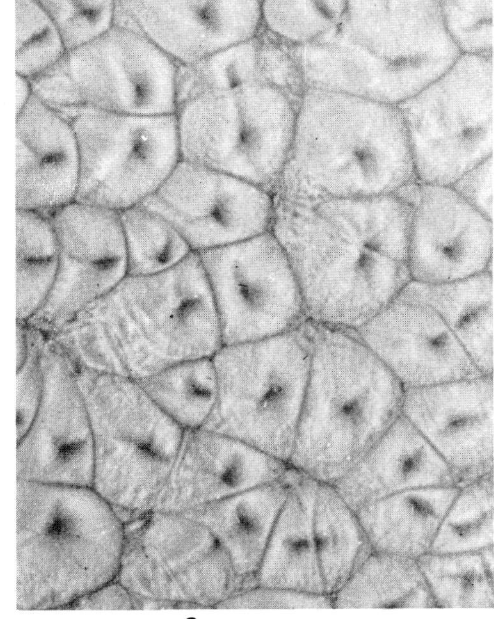

7.2.6 a

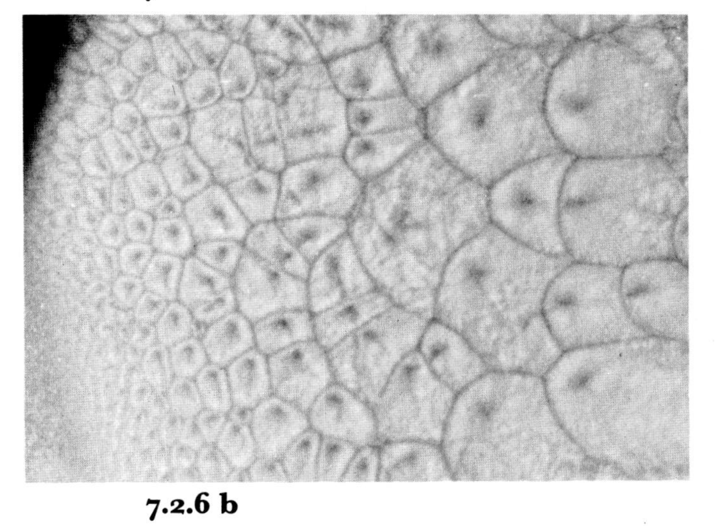

7.2.6 b

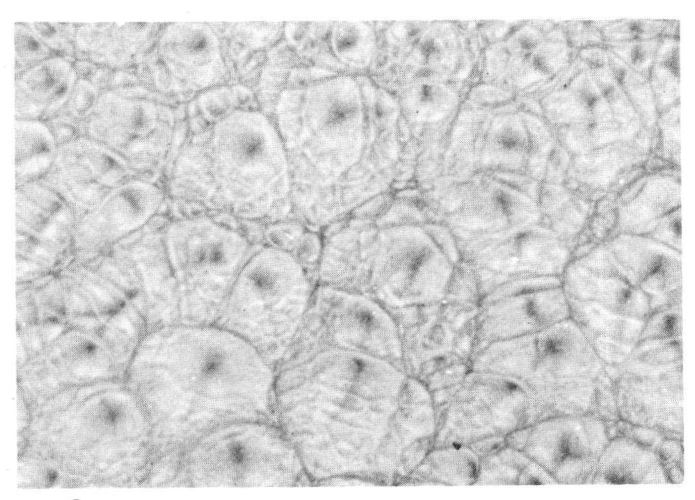

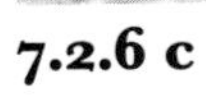

7.2.6 c

## 7.3 Radiative Effects

In the maintenance of cloud layers three factors operate to which there is no analogy in model experiments which themselves have a viscosity to which the cloud has no analogy. These are a continual subsidence of the air mass, the continued evaporation of water from the sea, and the continual loss of heat from the cloud top.

**7.3.1** This sea of cloud lies among the northern Pyrenees and is primarily maintained by radiation.

**7.3.2** By the time cold air which was perhaps 8,000 metres deep over Ireland has reached the Canary Islands shown here it has subsided to about 1,500 metres. The air above is of more westerly origin. In spite of warming and drying caused by this subsidence the air remains cool enough to contain cloud because of the heat loss by radiation, and damp enough because of the evaporation. The cloud has by this time become a layer very like altocumulus when seen from above. Even sea fog often has a similar appearance because it too has a motion dominated by the heat loss from the top. (See 1.6.2, 1.6.8 and 5.7.12). In this picture we see the coast of Tenerife through a gap in the clouds looking south west from the north east end of the Island.

Above is cirrus remaining from the jet stream clouds in air of more westerly origin which is the warm air mass at a cold front and has subsided less than the cold air beneath.

**7.3.3** When a haze is rather dense and has a well defined top, radiation from the top of it can sometimes produce enough cooling to cause the condensation of cloud. This example shows cloud just beginning to form at around 3,500m at Chittagong at sunset in February. During the day haze had been convected up to this height over the hot dusty ground over the whole of Bengal, Assam and Western Burma. It is common for small waves which occur as a result of the ground cooling (see 5.7.11) to assist in the formation of cloud at a haze top, particularly when it is at around 1,000m or less, but in this case the winds were very light and the air not stable up to 3,500m so that waves at that level would be unlikely.

## 7.4 Glaciation of Altocumulus

When anvil clouds are glaciated the subsequent behaviour depends upon the proportion of particles frozen at the beginning. If only a few are frozen they grow rapidly in the presence of supercooled water droplets and fall out. If all are rapidly frozen the particle size may still remain so small that they have negligible fall speed and remain suspended as a cloud. In altocumulus there are not generally large droplets among numerous small ones such as are carried up from below in a large cumulus (see, for example, 2.1.3, 3.6.6 and 4.2.2).

At a temperature of −35°C any glaciation would affect almost all the particles very quickly (see 11.4) and little fallout would be observed. At −20°C a few only might be glaciated and these would fall out. At even warmer temperatures (say −10°C) glaciation is generally rather unlikely.

When glaciation and fallout occur, water is removed from the air containing the cloud so that the remaining water cloud is likely to evaporate rather quickly. The following pictures illustrate three cases where the ice particles fall out.

**7.4.1** These are not castellatus so that there are not strong upcurrents and the mixing with the surrounding air and consequent evaporation is slow. Consequently there is time for frozen particles to grow large enough to fall out before the parent cloud disappears. These clouds over the northern Appenines are probably around −20°C.

**7.4.3** The first impression on this occasion in January at Karachi was that there was a dense haze below the cloud made visible by the sun's rays across the cloud edge. Closer examination showed the falling ice particles in clusters from where the altocumulus had been glaciated. The glaciation had caused the holes. The holes do not have the appearance of fallstreak holes (see 4.3.5) nor of progressively glaciated layer cloud (see 2.1.3) which suggests that the glaciation was spontaneous and more like 7.4.1. What we see here is the evaporating fragments of water cloud in the hole and around its edges.

**7.4.2** Glaciation is very rapid in these rather small altocumulus over the Nevada desert at about 6,000 metres above the sea. The copious fallout present after the evaporation of the water cloud suggests that there might have been some natural seeding of the clouds with mineral ice-forming nuclei from the desert itself. This picture was taken at 9 a.m. in June and convection from the ground had not reached this level, but the dust could have been carried up the previous day.

**7.4.4** This looks like a hole formed by the natural glaciation of a layer of water cloud. Once nucleated the cloud is like that shown in 4.3.5.

7·3·1

7·3·2

7·3·3

7·4·1

7·4·2

7·4·3

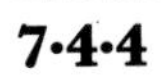

7·4·4

# 8 WARM SECTOR CLOUD

## 8.1 Sea Fog

In this chapter we study clouds formed as a result of the transport of warm moist air to a cooler region, and we are concerned only with the very low clouds, not with any role played by warm sector air in a cyclone, although that is where these clouds are seen. Warm sector is thus interpreted to mean warm moist air mass.

**8.1.1** Ocean liners of the North Atlantic very often have to travel through warm air masses containing cloud on the sea surface through which the sun is often visible. It is very difficult from the air to distinguish such cloud from layers at a considerable height above the sea because, once the cloud has become established by cooling at the sea surface, internal motion similar to that in altocumulus is induced by the radiation from the cloud top. This scene is typical of many an eastbound passage in which the ship remains in the same air mass for two or three days.

**8.1.2** Oceanic islands often become shrouded in cloud in a warm air mass even when there is no sea fog, either because of the small amount of lifting required to condense the cloud or because the radiation from the ground cools it to below the sea temperature and the increased mixing induced by the ground roughness cools a deeper layer of air. Often clouds of this kind are formed in a belt around a mountainous island, and they are then called Pollamjorki in the Faroes, where this picture was taken.

**8.1.3** Sea fog is common in June and July around the west and south coasts of Britain and similarly situated lands because the sea is still cool enough. The fog penetrates inland at night, but is dispersed by day, to return quickly across the coast in the evening. Even in bright sunshine the heating of the fog is negligible compared with the heat loss by radiation from its top, and the sea temperature is subject to very small diurnal variations except over tidal beaches.

This fog had penetrated into the Dovey estuary (central Wales) and was being converted first into cumulus (see 4.1.5) as soon as it moved over land, and then into anvil stratocumulus. The fog, with its small internal motions is distorted by the coastal shear, and has a very diffuse top compared with the fresh cumulus thermals beyond.

## 8.2 Low Stratus

Stratus is often a misnomer for the rather patchy cloud that is formed in the neighbourhood of hills. It is only correctly applied to layers of fog or of cloud not far above the ground or sea which are almost formless, and therefore not worth photographing, when seen from below. (See 1.3.3 and Chapter 9).

**8.2.1** When a warm air mass crosses an elevated coast (Cornwall in this case) the lifting begins a mile or two out to sea so that the cloud edge is formed there. The air had not been cooled enough to form cloud over the sea but needed to be lifted less than 100 metres in this case, which shows that the cooling by mixing from the bottom over the sea had not reached up to that height.

**8.2.2** Early in the day, before mountainsides have been much warmed by the sunshine, patches of cloud are formed in the rather stable moist layers coming in from the sea. Anabatic flow soon disperses these, and while doing so produces cumulus with rather variable base height. This is a view from Cader Idris (N. Wales).

Wave clouds, discernible below the cirrus over the Lleyn peninsular to the northwest, are a local sign of frontal activity.

**8.2.3** In among hills far from the sea the stable air of a warm sector may contain almost saturated air in which cloud is formed on hillsides. The motion often visible in these clouds may be very different from that of the clouds above which are not constrained by the stability to pass around mountains and through valleys. Separation is often observed on mountain sides at the top of these clouds. This one was observed near Trawsfynnydd in N. Wales.

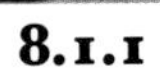

8.1.1

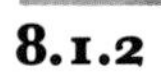

8.1.2

8.1.3

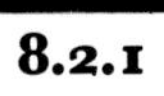

8.2.1

8.2.2

8.2.3

8.2.4

**8.2.4** Stable warm air masses penetrating into mountain valleys can often fill them with low cloud when that above is broken. This example was observed at Lauterbrunnen near the Jungfrau in April after a warm front with copious snow had passed. Such cloud is dispersed either by sunshine or by mixing with the drier air above, in which case it will behave like scud. At the time of this picture the wind was increasing and beginning to mix the stable air upwards, with some scud on the hillsides.

8.2.5

**8.2.5** Moist air masses in Antarctica readily produce stratus on mountainsides. The upflow may be produced either by direct impact of a strong airstream on a mountain or by the higher level wind being in the opposite direction causing the rise of the moist air up the lee slope. In this case the upper level cloud is moving away to the left while the low cloud ascends the lee slope (see 5.7.3 and 5.7.5).

The cloud has a definite condensation level which is clearly visible on the distant mountain side: this indicates that although there are clear skies above, cooling on the mountain side is not an important factor in the cloud formation.

Isolated steep mountainsides are often obscured by this kind of cloud while the surrounding sea is clear. The Rock of Gibraltar is often shrouded in cloud in an east wind when a moist rather stable air mass fills the western Mediterranean. The Island of Maui (9.2.4) and other islands of the trade wind belt often have their tops shrouded in drizzly stratus.

## 8.3 Damp Air and Ground

Most sea fog is rather old cloud so that it has probably been producing a small amount of drizzle for a long time. When it is lifted over a coast, there are plenty of larger cloud droplets ready to grow rapidly because the amount of liquid water in the cloud is increased. Copious drizzle is therefore often deposited on high coastal ground. During the morning in the summer the sunshine warms the ground through the fog sufficiently for it to be evaporated, whereupon the wet ground begins to produce a steaming fog (see 9.4).

The steaming fog evaporates as it mixes upwards into the air above and produces first scud and then ordinary cumulus. The following pictures show effects of illumination which have to be taken into account when photographing these phenomena.

**8.3.1a** The scene is near Aberystwyth in a westwind in April. The warm sector cloud has cleared to let through the sun at about an hour after noon. The ploughed field on which drizzle has fallen is quickly warmed and the surface begins to steam. The steaming fog (see 9.4) is contrasted with the shadows on the rough ground in this view towards the southeast.

**8.3.1 a**

**8.3.1b** Because of the great range of brightness it is necessary to give one quarter of the exposure used above to show the cloud forms. The nearest cloud is scud which is rapidly evaporating as it forms: nowhere is it dense so that much light is transmitted through it. Beyond it, another mile or two inland, there is ordinary cumulus which has sharply outlined brightly reflecting tops.

**8.3.1 b**

**8.3.1c** When the illumination is more nearly from behind the observer, as in this view towards the NNE, the steaming fog which produces only a small amount of backward scattering is no longer bright compared with the ground on which no shadows are seen. The dark scud, however, still contrasts well with the higher white cumulus beyond.

Steaming fog is always best seen looking towards the source of light. It is often seen on roads in sunshine after a shower.

**8.3.1 c**

# 9 FOG AND INVERSIONS

## 9.1 Radiation Fog

Radiation fog is the result of the cooling of the air by contact with and proximity to ground which has cooled by long wave radiation, mainly in wavelengths to which the water vapour in the atmosphere is transparent. In order to cool the air by contact some stirring due to motion over the ground is required, and this sometimes occurs after sunrise following a calm night. On the other hand there may be enough motion to form the fog before dusk. Nocturnal fog is not illustrated, although fog is probably a more common phenomenon at night.

Cooling by proximity to the ground is due to the loss of heat by radiation from the lowest two or three metres of air in wavelengths which are absorbed and emitted by water vapour. This is probably the cause of very low, calm, mist patches.

When the fog is very shallow so that trees can be seen in it from the air, the radiative loss from the ground is only partially prevented, so that structure does not always appear. But deep fog soon acquires a structure like altocumulus (see 7.3) when the radiative heat loss is mainly concentrated at the fog top.

**9.1.1** Although the plume from this Essex cement works can be seen in stereo to be well above the fog, the fog is mostly thick enough (about 100 ft.) to have structure. Some of this structure is due to wave motions as it drains down unlevel ground. Fog is usually eroded at its edges because sunshine warms the ground beneath it most where it is thin.

**9.1.2** On the same morning as 9.1.1 the fog near London Airport was much shallower over the slightly higher ground surrounding Windsor Castle. The fog has very little structure.

**9.1.4a** Mountainside fog just after sunset in calm air. Once the fog has formed by contact with the ground the air begins to be cooled directly by radiation. The mountain is about 350 metres high, in the foothills of Cader Idris (N Wales).

**9.1.5** It is not only in hollows that radiation fog is formed! There was a wind up this slope and the uniformity of the condensation level shows this. On the other hand only on the cooled mountain surface (Craig cwm Llwyd, Cader range) is cloud formed (compare 8.2.5 for example). The low evening sun shines on the mountain beneath the upper cloud whose base is dark.

**9.1.3** Even over the desert of Iraq fog can be seen in the Euphrates Valley on a December morning. At two hours after sunrise the absence of heating of the river and the fact that the fog was denser in the valley than close by are the causes of its persistence. Snow had fallen the previous night in the Persian desert, and this was a cold air mass freshly arrived from the north.

**9.1.4b** Three minutes after 9.4.1a the cloud has been carried down the slopes by katabatic flow and some of the cloud has been evaporated by the subsidence.

## 9.2 Valley Inversions

In this context an inversion means a sharply defined stable layer at the top of air which may be stable or mixed according to circumstances. (See diagram 9.2.i on page 94.)

**9.2.1** Looking towards the afternoon sun in December west of Geneva the haze top among the Jura Mountains shines brightly. From 11,000 metres the distant horizon is the haze top, which covered much of France. Although the air was stagnant in an anticyclone there is motion where a cloud lies under the inversion because of the downward convection. The air beneath the cloud is well stirred.

9.1.1

9.1.2

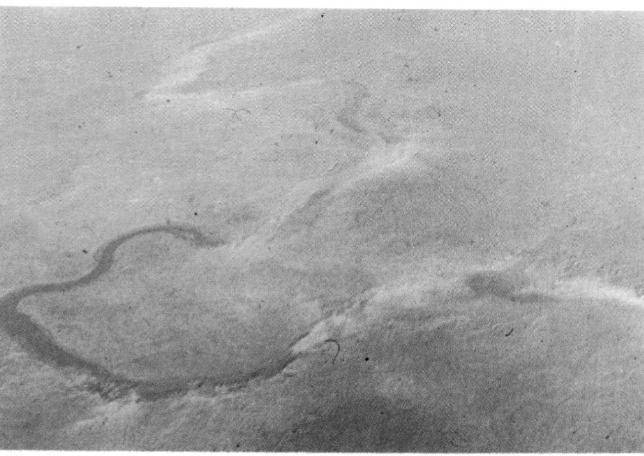

9.1.3

9.1.4 a

9.1.4 b

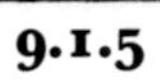

9.1.5

9.2.1

**9.2.2** The valleys surrounding Milesovka (835 m; 60 km. NW of Prague) are often full of pollution from generating stations and other industries. On this occasion billows occurred on the top of the cloud which lay under the inversion at the top of the pollution. The shear was between the stagnant valley air and the wind moving more freely above.

**9.2.4** One of the most persistent inversions in the world is found in the trade winds. It rises as the air moves eastwards, and is therefore higher here in Hawaii, for example, than in the Canary and Cape Verde Islands (7.3.2). The crater of Haleakala (on Maui) lies at about 3,000 m. with the highest point of the rim about 300 metres above the lowest. The trade wind inversion is just below the mountain top. During the day the anabatic flow causes the cloud below the inversion to ascend the final slopes and invade the crater, ascend the slopes of its peripheral mountain wall, and into the very dry air above the inversion and be rapidly evaporated.

**9.2.6** In some areas such as Iceland in summer, where, because of the long days convection is rather powerful, cumulus clouds form in a ring at the snow line around the larger mountain peaks. This is a view of Mont Blanc from the Col des Aravis. There is an inversion at the snow line with occasional cumulus rising through it from the highest mountains which do not penetrate it. The cloud at the top of the picture, like that below the snow line in front of the mountain, is essentially anvil stratocumulus.

**9.2.7** The cooling effect of cloud causes the air in the cloudy valley to become colder than in the adjacent cloud-free valley. In the morning the land under the cloud warms more slowly than in full sunshine, and the temperature difference is further accentuated. Consequently the cold air flows through the gaps in the mountains and as it spills over the floor of the warm valley the cloud in it is dissipated by convection which gives it a lumpy structure (see 9.3.3). The formation of the cloud in this valley in Pennsylvania was at least partly caused by the radiation from some rather persistent plumes from paper mills (see 10.2.2).

**9.2.3** Over the southern Red Sea area there is an inversion almost all the time. The highlands of Eritrea 2,000–3,000 m. are usually above it so that to the east of Asmara this is a typical morning scene. The cloud is evaporated first from the slopes facing the SE but persists on the northern and western slopes, especially where they are steep. Usually most of the cloud disappears by day, but reforms at night. Anabatic winds blow up the slopes as far as the inversion. Most of the rain in this neighbourhood is from convection from cumulus starting above or penetrating the inversion.

**9.2.5** In a mountainous area (in this case, the French Alps near Chamonix) an anticyclonic inversion tends to become located at the snow line because this is the upper limit of anabatic flow. If the inversion is lower it is raised by the anabatic flow, if it is higher it is lowered by subsidence. Often, as in 9.2.1, cloud is formed at the inversion, and this tends to perpetuate it by radiation from the cloud top. The inversion also becomes a marked haze top because this is the upper limit of anabatic convection. The cloud in this picture is formed on the north side of the valley because the mountain face is directed towards the sun. This contrasts with 9.2.3 where the sun is evaporating the cloud.

## 9.3 Sea Fog

Much sea fog is warm sector fog. In this section we are more concerned with the behaviour of fog over relatively small areas of cool sea.

**9.3.1** Fair Isle is about half way from the Orkney to the Shetland Islands, and it often lies on the boundary between North Sea water and the Gulf Stream, which is much warmer. In this view we see the harbour which lies on a narrow neck of land in the middle of the picture, engulfed in fog which covers the sea beyond, but not the water on the near side.

**9.3.2** Sea fog often exists near coasts because it has been formed over land at night and moves over the sea in a katabatic wind, and remains unevaporated for many hours of daylight. This is common in the Alboran Channel (Western Mediterranean). Sea breezes carry such fog inland some distance in the evening in particular when the ground cools so that it is not evaporated as it crosses the coast; but it may be lifted off the ground at that time by the first onset of the katabatic wind. Thus the air below the fog on the hillside across this bay in S. Greenland may be katabatically warmed air which undercut the sea fog at night. It is rising off the hillside where anabatic flow has just begun.

9.2.2

9.2.3

9.2.4

9.2.5

9.2.6

9.2.7

9.3.1

9.3.2

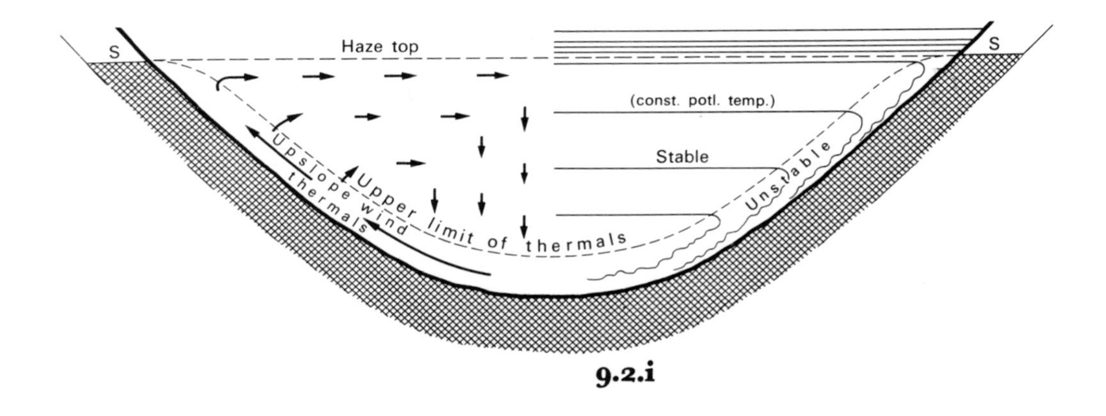


9.2.i

**9.2.i** In the right half of the diagram the surfaces of constant potential temperature are shown with a large stable gradient at a haze top inversion. Below it is less stable, but stable enough to cause the heated air to travel up the hillsides in a shallow layer.

When the valley is wider the thermals in the middle often reach up to the main inversion, and when cumulus are formed over the mountains they do not penetrate beyond the sub-cloud stable layer (1.1.iv, page 21) which becomes a haze top.

The upper limit of anabatic winds is either a very marked subsidence inversion, as in 9.2.1 and 9.2.3, or the mountain tops, or the snow line as in 9.2.5 and 9.2.6.

When the air over the low ground is stably stratified the anabatic wind is often only 100 metres or less in thickness on the slopes and contains updrafts in which gliders can be soared.

Katabatic winds are much shallower and are accompanied by almost no vertical motion in the main air mass.

**9.3.3** The sea fog shown in these two pictures lay in a band up to 16 miles wide off the coast north of Eureka, California. They were taken just before noon in August, when the warmth of the land was evaporating it.

The air in and beneath the cloud is kept cold by radiation from the cloud top. Convection from the sea causes a cellular (in the bottom of b) or banded structure in the cloud. The cold air spreads out to sea and by day inland also and the cloud is evaporated on both edges. The band of cold sea is produced by upwelling which is caused by the wind over a larger area of ocean.

The fog extends from the sea surface up to 400 ft. The thermals formed inland reach up to 500 ft., the land being nearly 200 ft. above the sea. There is a rise of temperature in the air above the fog up to about 1,500 ft. where it is 11°C warmer and much drier than at the top of the fog. A subsidence inversion is thus greatly reinforced at the bottom by the cooling at the fog top. In (a) the change in structure due to warming over the land is clear. In (b) the seaward edge is seen to be in street form as the cloud disperses, very much like the structure in the bottom right of (a).

In (b) a distant smoke plume can be seen carried up a valley by the sea breeze in the stable air.

**9.3.4** When a sea breeze front penetrates inland it often brings with it cool moist air, sometimes containing stratus cloud, or forming stratus cloud when it extends up hillsides. In this picture the cold air behind such a front is seen spilling into the Mackenzie Basin (about 60 miles inland in New Zealand, South Island) in early afternoon. The cloud was evaporated as it crossed the warm dry valley (rather like the cloud in 9.2.4).

Sea breeze fronts are common in many countries, and are most pronounced when their advance is opposed by a rather feeble wind (geostrophic). They are often clearly seen by short wave radar, though it is not known whether this is because of the sharpness of the humidity discontinuity or because of the insects (and birds) lifted at the front.

9·3·3 a

9·3·3 b

9·3·4

**9.3.5** In August sea fog is common in the Channel, though less so than in July. This is a view eastwards and the wave clouds are over the Cotentin peninsular south of Cherbourg about 35 miles away. There were patches of sea clear of fog, on the edges of which the fog was being slowly evaporated by upward mixing.

### 9.5 The Tropopause

**9.5.1** The tropopause in a cold air mass behind a cyclone is usually a rather sharp inversion with a very dark blue sky above. Below, the moisture reddens the sky as the sun sets. Here an aircraft is making a turn over the English Channel, over which there are small cumulus and a bright patch of light where the specular reflection is occurring. On either side of the sun is a mock sun at 22°. These are very commonly seen in high level ice clouds when the sun is low and the observer at the level of the clouds.

Other pictures of the tropopause are 4.2.6, 6.2.5 and 11.3.3.

### 9.4 Steaming Fog

**9.4.1** The air close to a hot wet surface quickly becomes saturated. At the same time its buoyancy carries it upwards to mix with cooler air above and condensation immediately occurs by the earliest mechanism envisaged by physicists, namely the mixing of two not quite saturated parcels of air of different temperature to form an oversaturated mixture. The excess water condenses as cloud.

Thermals can be seen rising in the steam.

Breath forms visible steam in cold weather and wet roads steam in sunshine and even horses after violent exercise do likewise. No other natural clouds are formed by this mechanism although it is important in the formation of exhaust trails and wet chimney plumes (see Chapters 10 and 11).

This shows Lac Genin (French Jura) on a calm August morning, a cold front having passed over the previous evening. The steaming continued for about 10 hours. (See also 8.3.)

# 10 WHITE PLUMES

### 10.1 Water-laden Plumes

White plumes may be artificial water clouds, clouds of hygroscopic particles, or clouds of solid white particles. It is not always possible to distinguish between these at sight. Particles of coloured material cannot form a pure white plume, although some plumes of slightly coloured particles appear white in certain lighting.

**10.1.2** Steam locomotives often emit steam laden plumes when at a standstill. A close examination of the plume shows that it is invisible close to the valve, but as soon as the hot water vapour mixes with cooler air condensation occurs. The heat released by condensation increases the buoyancy of the plume, but is lost when the droplets evaporate. In this scene (Moffat, Scotland) the plumes are rising to cloud level without evaporation because of the high ambient humidity.

**10.1.1** When the heat from the stack effluent is added to the moist air emerging from cooling towers the buoyancy is often enough to carry the plume up to natural cloud level. When the plume is vertical the dilution is a minimum so that evaporation may not be complete before the condensation level is reached. Cumulus clouds are then formed. This station is Ham's Hall near Coventry.

**10.1.3** The heat lost by evaporation of the plume from a steam locomotive in motion causes the plume to descend to the ground. The plume also contains flue gases. The original condensation is due in part to expansion, and is not the same as in a condensation trail where it is due to mixing alone.

9.3.5

9.4.1

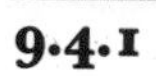

9.5.1

10.1.1

10.1.2

10.1.3

**10.1.4** Cement works plumes are usually white because of the condensation of the water evaporated from the slurry fed into the kiln. The condensation sometimes occurs inside the chimney, but occasionally a clear gap (see 10.1.5) exists between the white plume and the chimney top showing that the condensation occurs after efflux. When the plume evaporates a residue of whitish dust makes the plume visible for some further distance.

**10.1.6** The water is introduced into this plume at Bankside power station, London, during a washing process which removes almost all the sulphur dioxide. The plume is sharply outlined but the bright whiteness disappears as the water droplets evaporate into the drier ambient air. In the final stages of the washing, water previously evaporated into the plume is condensed as a cloud: if the ambient air is warmer than about 20°C the evaporation of this water almost always causes the plume to acquire negative buoyancy and descend to the ground. Particles remaining in the plume give it a white or bluish colour, and these are gradually dispersed. This view is from St. Paul's Cathedral at about chimney-top level.

**10.1.5** It is unusual for the water vapour in a power station plume to be sufficient to form a condensation trail after emergence: it can only happen with the fuels normally used if the ambient temperature is unusually cold. The ambient humidity does not make much difference to the formation, although it does affect the subsequent evaporation. In this example (42nd St., E. Side, New York) on a cold February morning the plume from the right stack produces condensation after efflux. The evaporation occurs rather rapidly because of the low ambient relative humidity. The cloud has the same appearance as any ordinary water cloud and the rapid evaporation and sharp edges makes it easily distinguishable from a non-evaporating cloud.

**10.2.2** The persistent plumes from these pulp mills at Tyrone (Pennsylvania) are of a hygroscopic salt emitted from the process (see 10.2.3). The nearest chimney is emitting a dustladen plume from a steam raising plant and the dense white plumes, which are from small cooling towers, soon evaporate. The effluent is trapped under an inversion below the mountain tops (Bald Eagle Mountain), which is common in this region. On the occasion of 9.2.7 cloud was formed in this valley, but not in the neighbouring one, because of the heat lost by radiation from these plumes.

## 10.2 Other White Plumes

**10.2.1** The middle chimney (Liverpool) is emitting a plume containing a larger proportion of sulphuric acid particles than is usual from power stations. The causes are not properly understood, but it does depend upon furnace, boiler, and flue design. This plume is gradually diluted and shows no sudden evaporation like a water droplet cloud (10.1.5). The plume on the left is composed mainly of fly-ash particles from a pulverised coal boiler. The other is a chain grate stoker.

Similar white acid plumes are often seen from oil burning plants.

**10.2.3** These pulp mills near Seattle emit a plume which is persistent because of the sodium sulphite particles which do not evaporate. The absence of evaporation makes it possible for no mixing to occur in the stable turbulence-free air in the evening. The plume travels without further dilution for several miles.

**10.2.4** Air is bubbled through molten iron to remove impurities and many of these are removed from the plume by wet washing. On emergence from the chimney the plume is thought to contain large amounts of silica and possibly large amounts of liquid water, although no evaporation of it is seen in the picture, possibly because of the presence of hygroscopic nuclei.

**10.2.5** Crepuscular (=evening) rays are shadows on smoke or other forms of haze, but are not only seen in the evening, although they are more easily seen then. Often forest fires cause such dense patches of smoke that they are not easy to distinguish from cloud. They do appear browner or redder than natural clouds.

This cloud in the upper right part of this picture is smoke below 2,000 metres, but is often reported as cirrus.

10.1.4

10.1.5

10.1.6

10.2.2

10.2.1

10.2.4

10.2.3

10.2.5

# 11 CONDENSATION TRAILS

Condensation trails, usually called contrails, are water clouds formed by the air motion produced by an aircraft or by the condensation of the water vapour in the exhaust. In principle the second mechanism is the same as that illustrated in 10.1.5, but the shape of the trail is very much affected by the motion left by the aircraft.

## 11.1 Cloud close to the aircraft –

**11.1.1** On a day of high humidity near the ground, such as when the ground is very wet, a cloud may be formed in the low pressure region above an aircraft wing. Since the air passes through this region in about 0·1 sec. the condensation and evaporation must occur in a time of the order of 0·01 sec. This phenomenon is most commonly seen at the moment when the nose of a jet aircraft is pointed steeply upwards just after take-off. The steepness of ascent of this Comet IV at London Airport is seen from the inclination of the horizon.

There is also a trail of particles in the wing tip vortex, but this is not a cloud of condensation: it is a cloud of spray from the fuel tank vent and the particles are large enough to be centrifuged out of the vortex core, which therefore appears darker.

**11.1.2** Another place where the relative humidity is high is just below the condensation level. There is condensation in the low pressure region at the centre of the vortex trailing from the tip of the lowered flap of this Viscount aircraft coming in to land at Dublin Airport.

**11.1.3** The blade of a propeller is an aerofoil, from the tip of which a vortex trails. The strength of the vortex depends on the thrust of the blade and this is largest when the aircraft is in a steep climb, as is usual after take-off. Near to cloud base, or pieces of scud, cloud is formed in the low pressure region at the core of each vortex. Each propeller trails four entwined vortices so that we see that the aircraft (Vickers Vanguard) travels about one engine length for each propeller rotation. The vortices are wrapped around the wing because the air from the propeller travels partly above and partly below it.

**11.1.4** When it first condenses the exhaust trail cloud is composed of water droplets, but in most conspicuous trails the particles are frozen, and there is only a small part of the trail, close to the aircraft, where there is evidence for spherical droplets: occasionally iridescence can be observed, as in this picture. Even when it occurs this is not easy to see and still less easy to photograph because of the general brightness of the sky close to the sun. This example was photographed through a 500mm lens, and the colours were only visible to the eye through the camera viewfinder (ie through the telescopic stopped down lens), and the shortness of the coloured part of the trail indicates that the cloud was freezing within about a second.

## 11.2 Exhaust trails in aircraft wake

**11.2.1** From very few modern aircraft can the wake of the aircraft itself be seen. This view from a 'Flying Fortress' (B17) shows the downwash of clear air into a cloud. In this case the accompanying fluid is partly clear and partly cloudy air, so that the gap in the cloud is probably filled in rapidly by cloud above the vortices moving inwards. To make a persistent clear lane the vortices must be formed in the clear air some distance (about a wingspan or more) above the cloud. (See 11.2.6 and 9, and 11.4.4.)

**11.2.2** As soon as the aircraft began its turn the downwash became tilted outwards, and the development of the blobs in relation to the top of the curtain of cloud shows this. This behaviour shows clearly that the buoyancy of the trail due to the heat of the exhaust has an effect which is negligible compared with the effects of the downward impulse imparted to the air by the aircraft and the buoyancy subsequently developed as a result of descent through stably stratified air.

**11.2.3** This trail was made by a fighter aircraft with engines placed centrally so that very little exhaust travelled round the vortices to be detrained at the top of the accompanying fluid (see p 102-3). The air was evidently rather stable because the blobs soon formed.

11.1.1

11.1.2

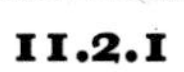

11.2.1

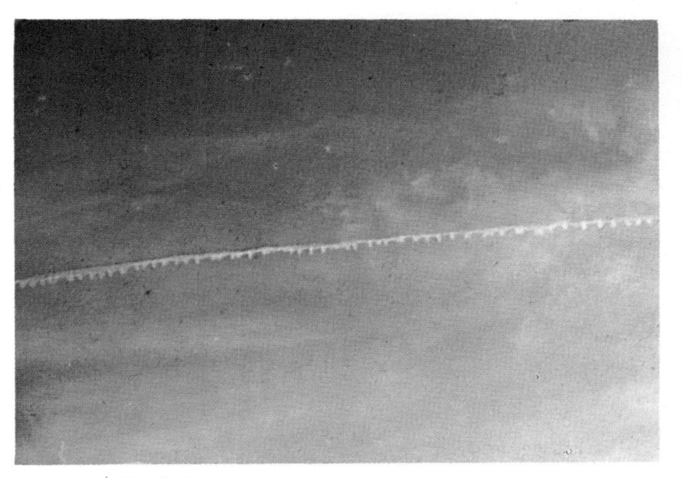

11.2.3

11.1.3

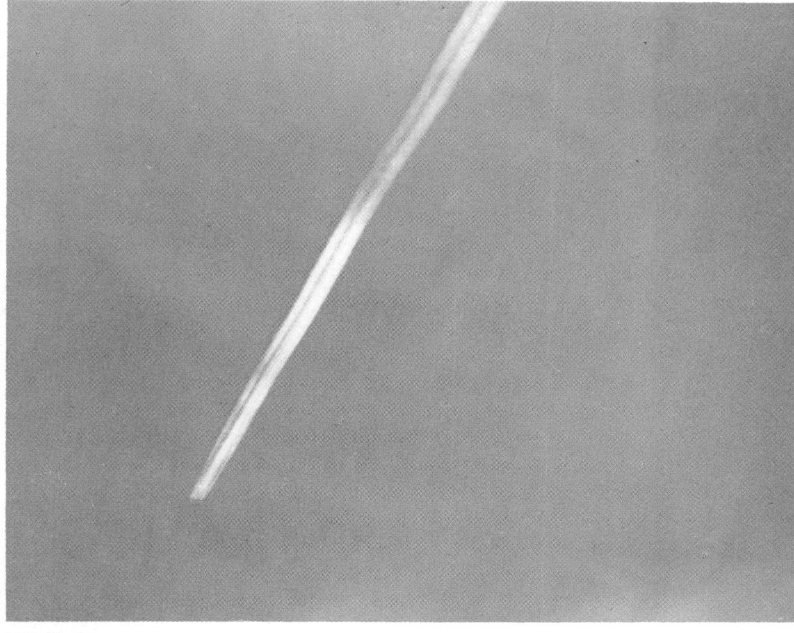

11.1.4

11.2.2

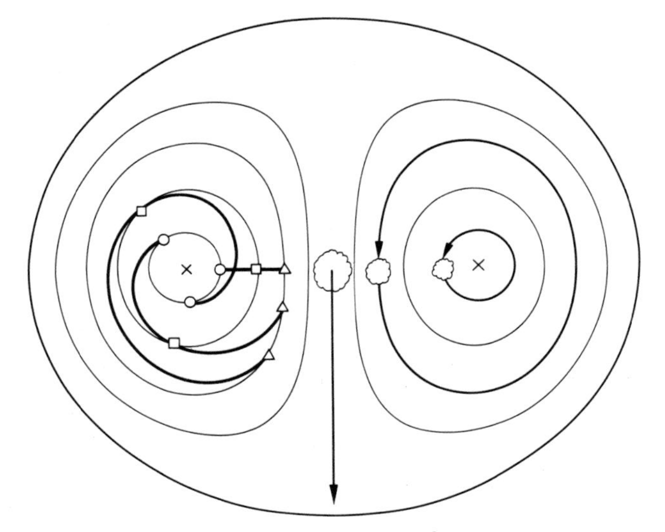

**11.2.i**

**11.2.i** The two crosses represent the axes of the vortices trailing from the wingtips of an aircraft. In this vertical section the closed curves represent tracks of the air circulating round the vortices. The outside curve encloses the "accompanying fluid", which is the name given to the air which travels downwards, with the vortices, through the surrounding air. Since the exhaust is emitted between the wingtips it is inside the accompanying fluid. In the right half of the diagram the heavy lines show how exhaust trails would circulate around the vortex. Exhaust emitted along the centre line would travel directly downwards, and would then be divided when it reached the bottom of the accompanying fluid. The motion becomes very much modified before such a division would occur.

In the left half we see three points on a horizontal line marked by a circle, square, and triangle. The circle travels round the vortex more rapidly so that three successive positions of the originally horizontal line of particles are shown. In particular, as viewed from a distance, the inner exhaust trail shown on the right would appear to travel round the outer one (the square moves to a position beneath the circle, and the triangle likewise but more slowly).

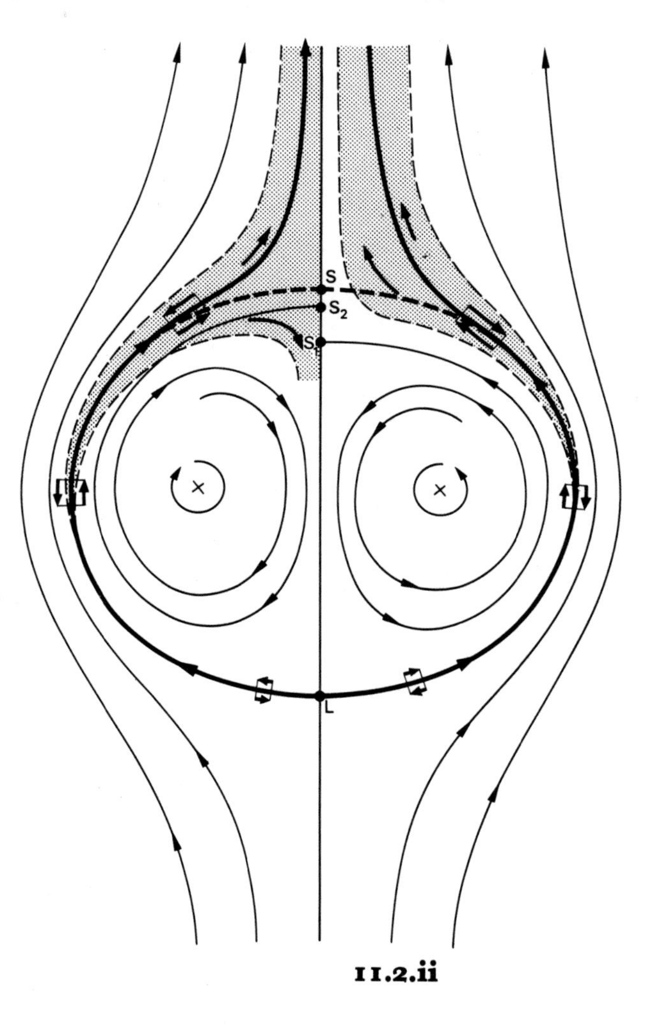




**11.2.ii**

**11.2.ii** The motion shown in 11.2.i does not persist because of the buoyancy acquired by the accompanying fluid as it descends through a stably stratified environment. As the air moves from the lower stagnation point L the juxtaposition of buoyant air, inside the boundary of the accompanying fluid (heavy line), with the air outside produces vorticity in the direction shown by the small pairs of arrows. This vorticity, which is in the opposite direction to the principal vortices, has the effect of moving the upper stagnation point from its original position S down to $S_2$ so that all the air between the streamline (left half of diag.) to $S_2$ and the outer boundary starting at L is "detrained". It is left behind so that the accompanying fluid becomes smaller. This has to happen because the buoyancy force acting on the accompanying fluid decreases the downward momentum of the system, and the only way this can happen is for the accompanying fluid to decrease in volume: this has the effect of bringing the two vortices closer together and since their circulations are not decreased their downward velocity is actually increased. This is more than offset by the decrease in volume so that the momentum still decreases.

The vortex sheet which is generated on the outer boundary of the accompanying fluid is unstable, and so a mixing zone will appear, which is represented by the shaded area which gradually widens upwards. To begin with the stagnation point $S_1$ is below this region and so all the mixed fluid is left behind; this is shown in the right half. But after a minute or so, in the case of typical values for airliners at about 11,000 metres, the stagnation point rises nearer to the original position. The consequences of this are illustrated in the left half of the diagram, with some of the mixed fluid travelling down the middle of the accompanying fluid. As soon as this penetrates beyond about the middle of the system the main vortices become unstable because the buoyancy of this mixed fluid will cause the generation of circulation, opposite to the principal vortices, inside the accompanying fluid in circuits surrounding the vortices. Mixing then very quickly penetrates into the cores of the vortices. The consequences of this sequence of events is illustrated in the next diagram and the pictures following.

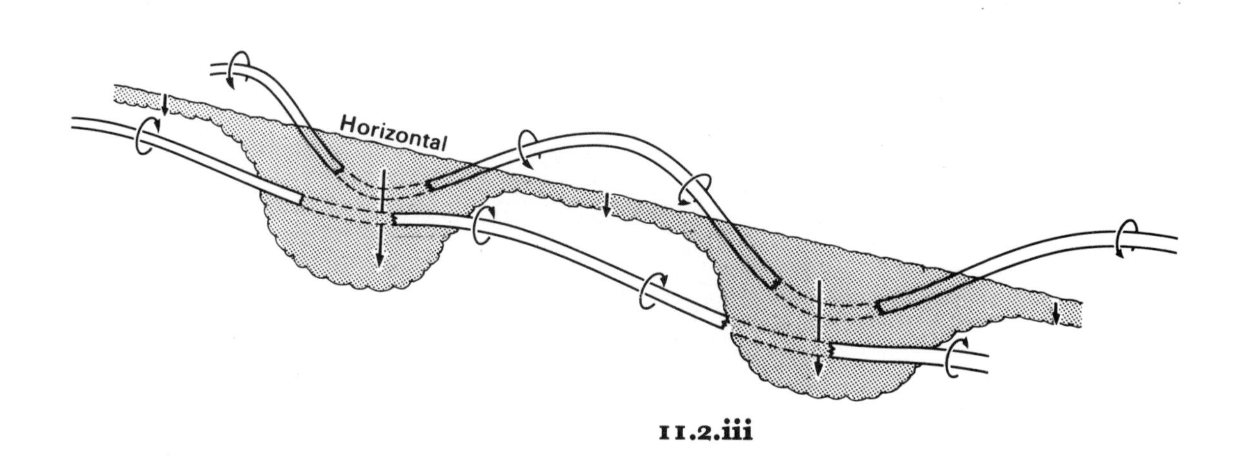


**11.2.iii**

**11.2.iii** This perspective view from an angle above of a vortex pair illustrates how irregularities are magnified because the buoyancy causes an increase in descent velocity which itself increases the buoyancy. The cloud carried down between the vortices descends in a series of blobs. The descent ends when the mixing penetrates into the vortex cores and the two loose ends of the vortices tend to join across the system to form a series of loops. The lowest parts of the vortices are closest together.

If cloud is present in the vortex cores it will remain until mixing reaches it, when it will rapidly disappear because of the mixing and the rapid rise in pressure as the rotation ceases.

**11.2.4** The contrail is formed about one wingspan behind the aircraft when the exhaust has mixed with sufficient air of the environment to produce saturation. The edges of the trails are ragged to begin with but further mixing evaporates these and the trail becomes smooth: this indicates that the condensation persists only in the low pressure region of the vortex cores.

Close to the aircraft it is seen that the two trails from the inner engines descend further than the outer ones, those on the right being separately visible but the outer left one being behind the inner left one. Often one complete rotation of the inner round the outer one can be seen.

In the second picture the trails appear hollow. This is because the exhaust did not enter the middle of the vortex core. The particles are too small for centrifuging to be significant.

The irregularities are intensified, and at the right end the mixing penetrates in a few places into the vortex cores and dissipates the cloud. The spacing of the loops must be determined by the distance apart of the vortices: irregularities of all wavelengths are unstable, but two loops very close together merge into one, while new loops can appear between those that are rather far apart. The interval between the first and last pictures is 90 sec.

Motion identical with that shown here has been observed in smoke trails emitted at the wing tips of a Comet aircraft at 3,000 metres which is far below the normal altitude for contrails.

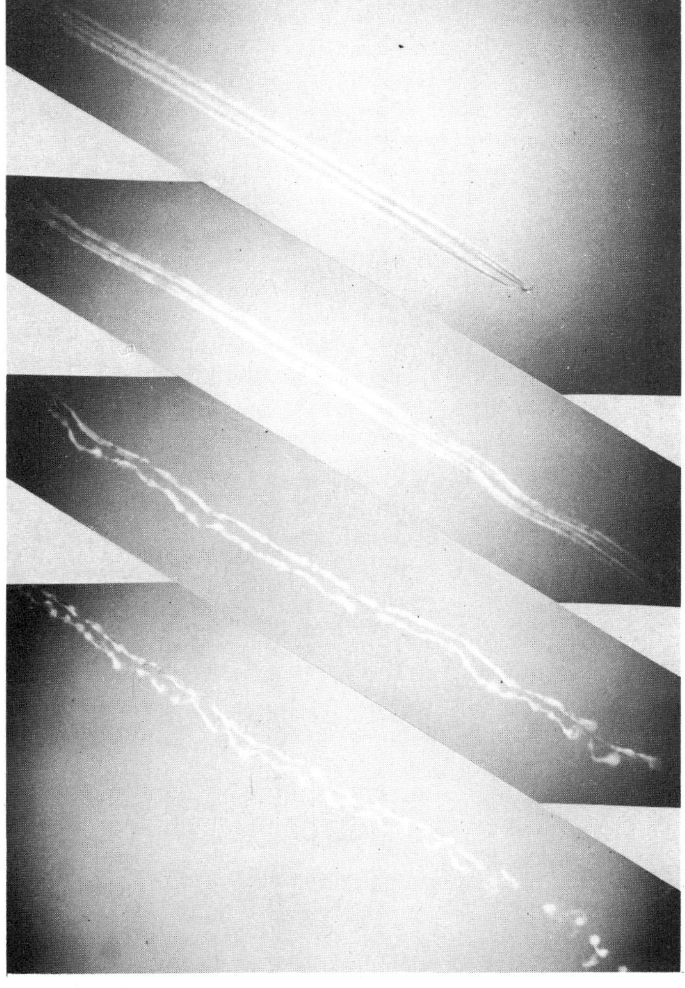

**11.2.4**

**11.2.5** The condensation outside the vortex cores persisted in the trail of the first aircraft to pass, but not in the trail of the second. It seems probable, therefore, that the first trail was frozen as a result of being at an environmental temperature near to or below −40°C. (See 11.4.)

These pictures were taken looking vertically upwards over a period of about three minutes, and because the aircraft was on a turn the downwash was inclined outwards. The effect is to give a view rather like the side view of a straight trail (see 11.2.3, 11.2.8). Where the vortices are closest together they have penetrated farthest, and the blobs correspond to the places where the mixing has penetrated into the vortex cores. In the lowest picture the linear clouds are composed of the detrained accompanying fluid.

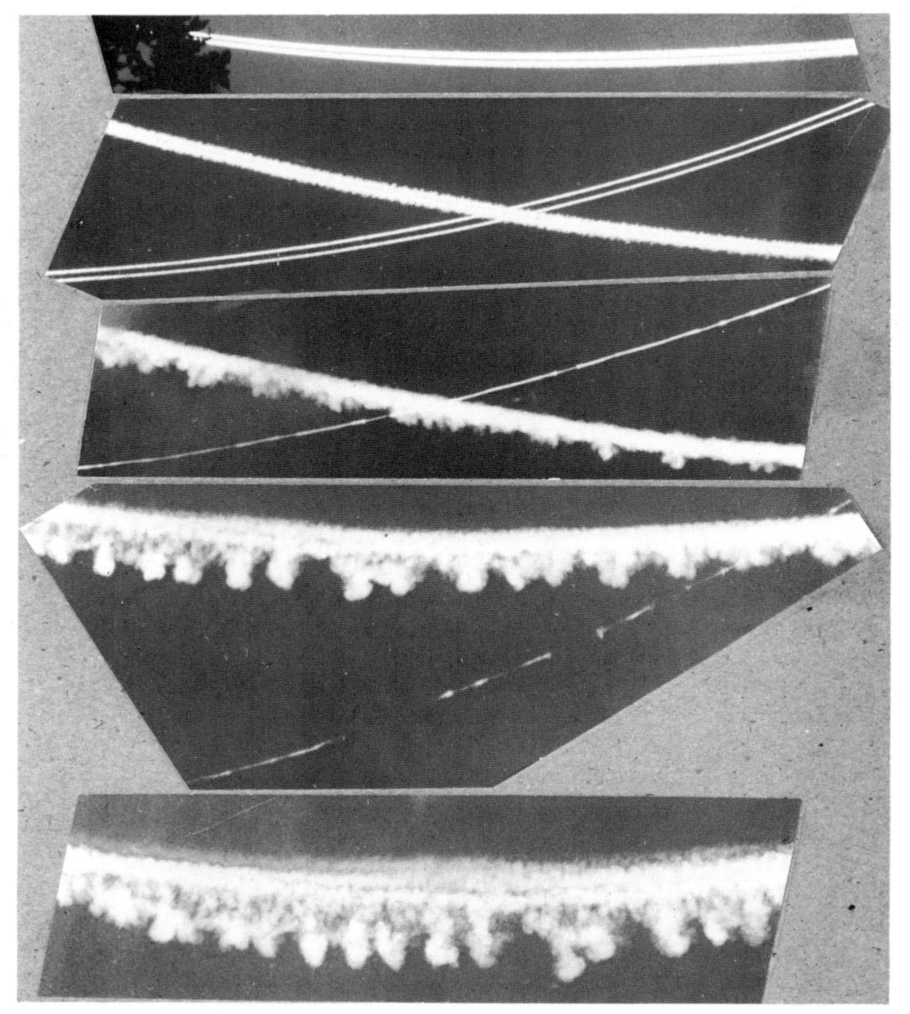

**11.2.5**

**11.2.6** The aircraft making this trail travelled from the lower left corner of picture *a* in a turn off the top of the picture. It seems that it travelled in the cloud for a short distance and that that part of the cloud was already frozen so that the exhaust quickly froze. The unusual feature was the absence of any trail elsewhere.

The distrail produced by the penetration of the downwash (see 11.2.9) into the cloud becomes progressively greater in *b* and *c* which were taken at one minute intervals. At the bottom of *c* and in *d*, which was taken of the earlier part of the trail two minutes after *c*, the penetration of the blobs of clear air from above the cloud is complete. The holes lasted another minute or so before they began to fill up.

**11.2.6 a**

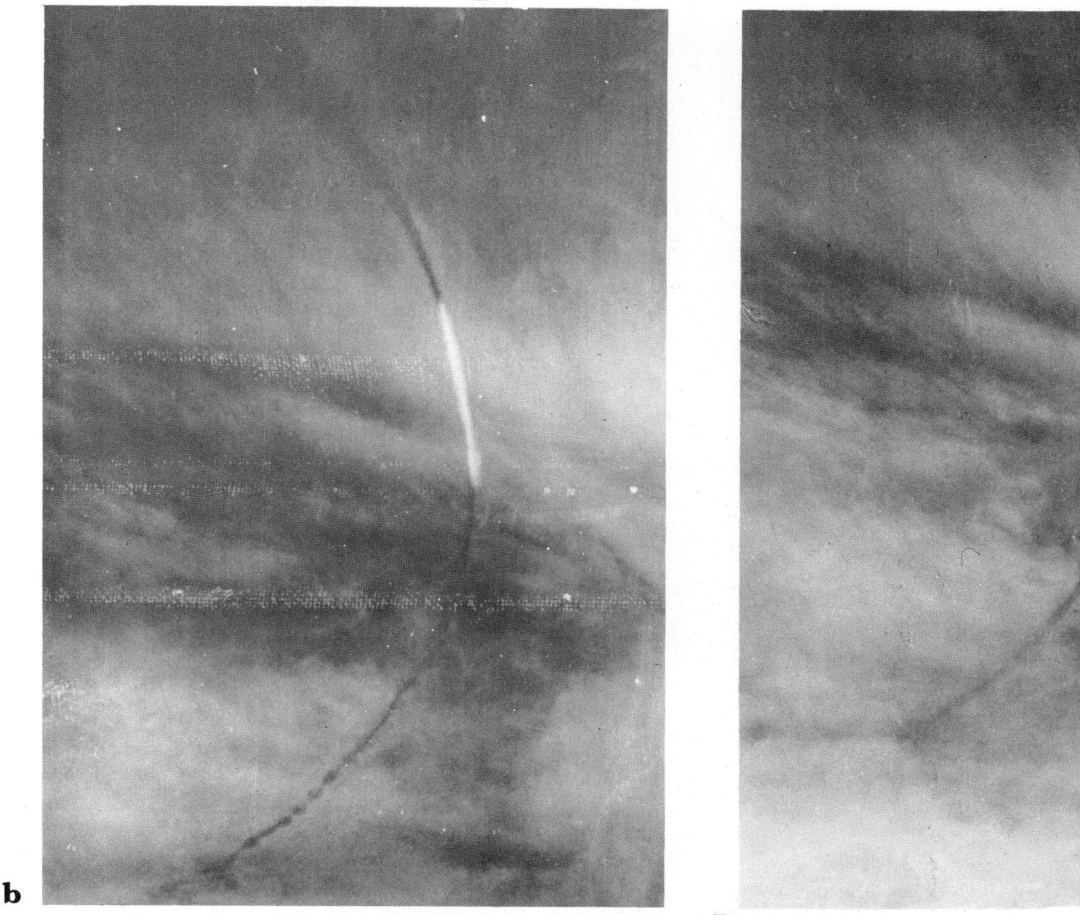

**11.2.6 b**

**11.2.6 c**

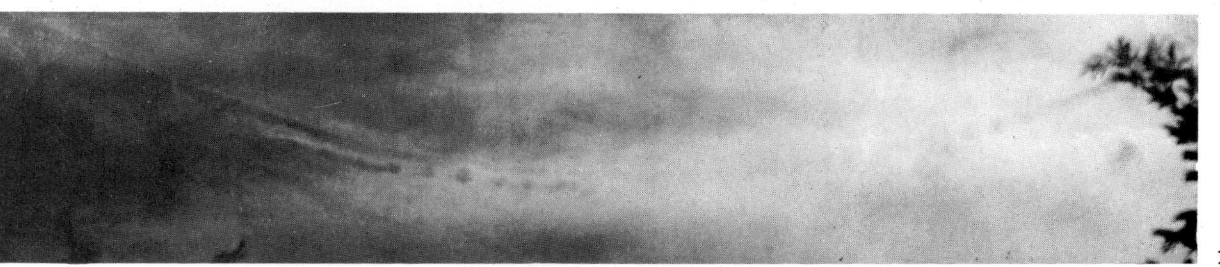

**11.2.6 d**

**11.2.7** Dense trails are formed from all four engines of this aircraft and by the time the trails from the inner engines have moved round to the outside they have begun to evaporate. The exhaust from the outer engines enters the cores of the vortices and the cloud persists in the low pressure regions only. There appears to have been some side slip because either the two vortices are unequal or, more probably, the amount of exhaust entering the vortes on the outside of the bend was less. This is similar to the feebler trail of 11.2.5.

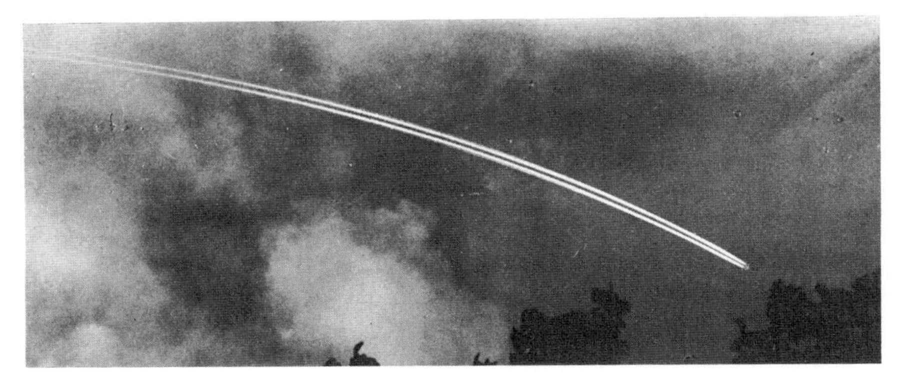

**11.2.7 a**

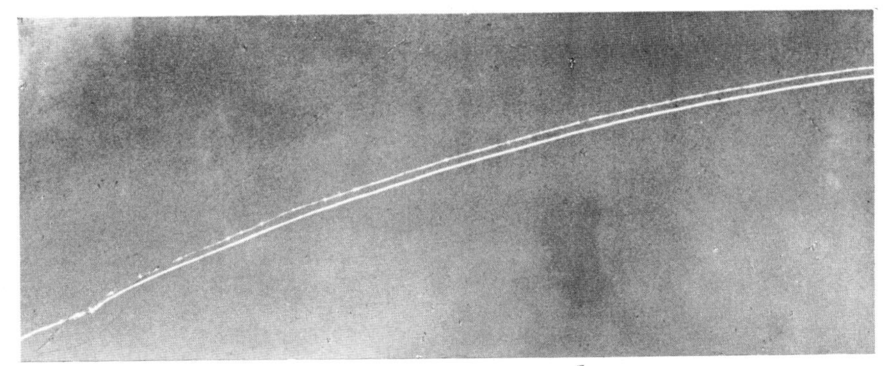

**11.2.7 b**

**11.2.8** The development of the curtain of cloud through detrainment of the accompanying fluid is seen in the development of this trail. The time interval between *a* and *b* is about the time taken for the aircraft to fly the length of trail shown, so that these pictures could be be placed end to end to give a representation of the trail growth.

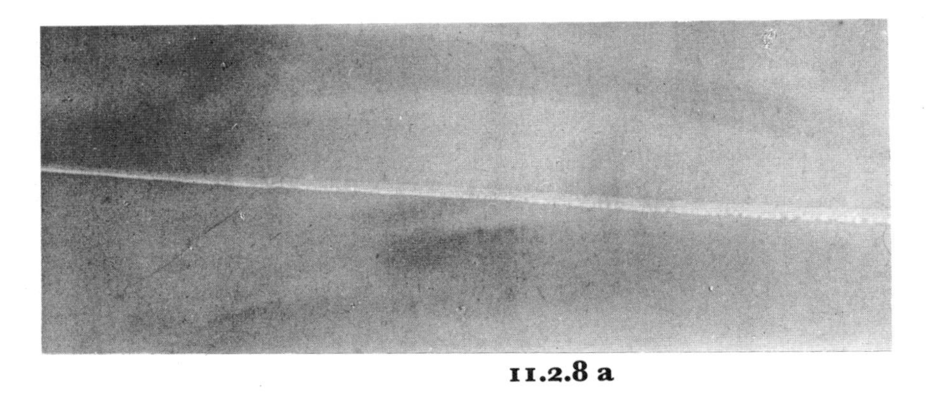

**11.2.8 a**

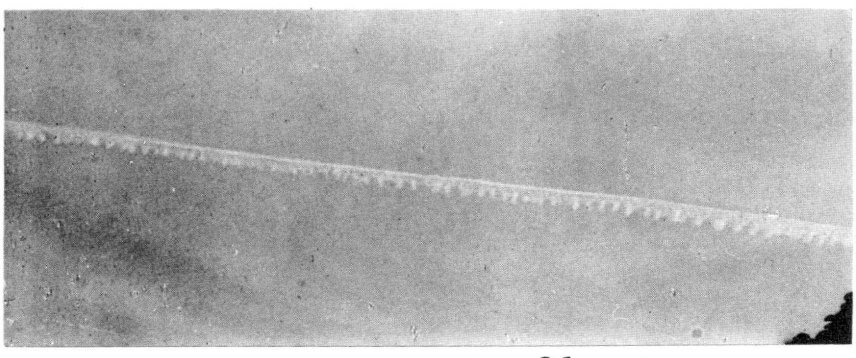



**11.2.8 b**

**11.2.iv** The physical basis of contrail formation is shown by these diagrams giving the saturation mixing ratio in gm. of water per kg. of air at three different pressures. The position of typical aircraft exhaust on the diagram is shown by the direction of the arrow, the temperature being a few hundred °C.

If the ambient air is at a point above the water saturation curve the excess water must be present as cloud. The mixing of the exhaust with increasing amounts of air produces a mixture whose point on the diagram moves (roughly parallel with the tangent through the exhaust point) towards the ambient air point.

If the ambient air point is in the shaded region cloud will form at some time because the mixture point will have to pass above the water saturation line. If the cloud does not freeze it will evaporate when the mixture point crosses the saturation line on approaching the ambient air point. If the cloud freezes and the ambient air point is in the doubly shaded region the cloud will not evaporate because the mixture will remain above the ice saturation line; the trail is then persistent.

Trails are rare at 550 mb because the ambient air point is usually well to the right of the −28°C line, and so saturation is never achieved in the mixture. At 400 mb trails occasionally occur. At 250 mb the trails are rather common and always freeze if they are formed; they persist if the ambient mixing ratio exceeds about 0·13 gm. water vapour per kg. of air.

Trails formed in wingtip vortex cores are at a lower temperature and pressure than the ambient air and may be formed at a lower altitude but do not persist unless an ordinary trail would also persist. Trails like 11.2.7 always evaporate quickly, while those with exterior cloud present usually persist.

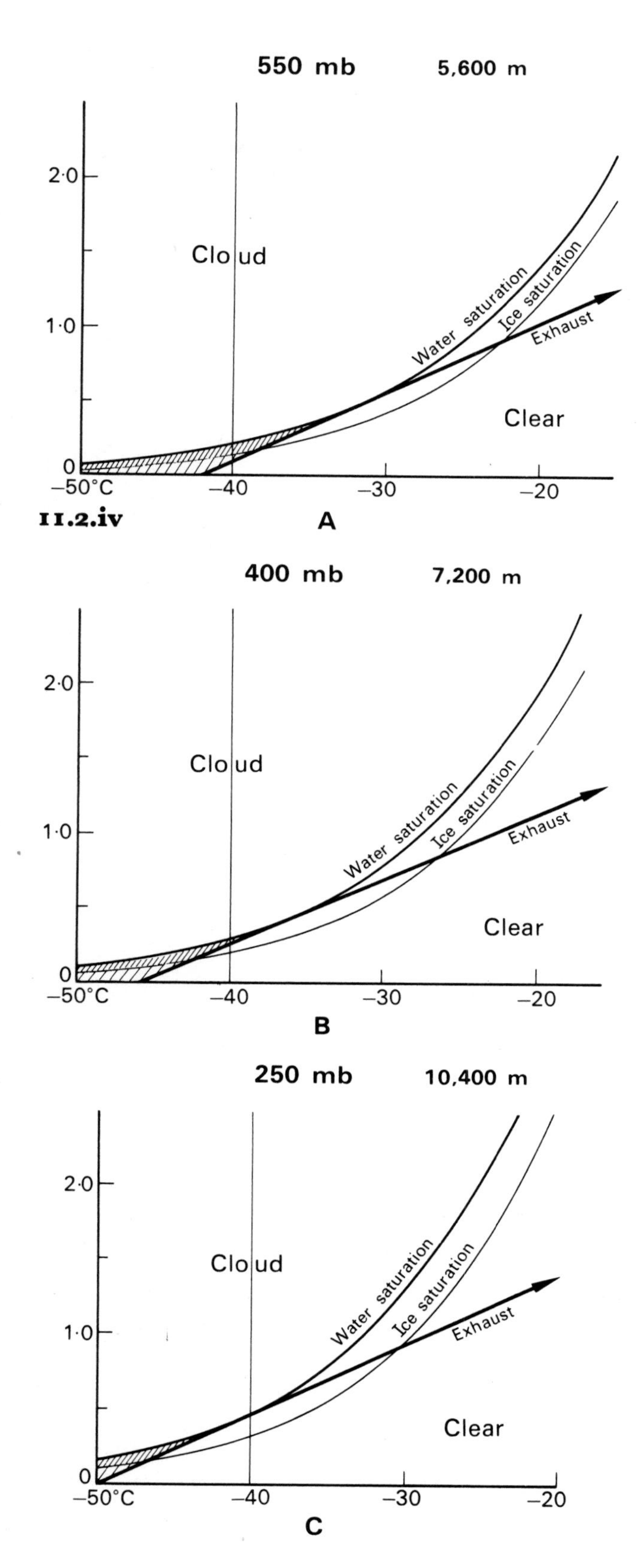


**11.2.iv**

**11.2.9** It is clearly possible for a distrail and a contrail to be produced simultaneously. In this case the contrail is denser than the very thin layer of cloud. The time was about sunset (in Pennsylvania) so that the contrail which was carried below the cloud by the downwash was better illuminated than the cloud which lay in the plane of the illuminating rays.

There is much uncertainty about the origin of the trail because usually the air above a thin layer of cloud is very much drier and unlikely to support a persistent trail. The trail appeared along only part of the length of this distrail and could have been formed of glaciated cloud (see 11.4.4).

## 11.3 Contrail Shadows

Shadows of contrails often produce dark strips on thin layers of cloud that are not easy to distinguish from distrails. Distrails, however, are rare in ice clouds because the clouds are seldom thin enough layers.

Some ice clouds are so tenuous as to have the appearance of haze, and when the observer is close to the plane of the sun and a contrail the shadow on the 'haze' is often very easy to see. By means of shadows the relative position of contrails and clouds may often be determined.

The dark shadow of a contrail is often very easily seen on haze when the observer looking upwards at the trail is close to where the shadow falls on the ground (ie the sun appears close to or behind the trail), see 11.3.3.

**11.3.1** Between the layers of a weak front contrails are easily formed. Here the contrail on the left casts a shadow on the thin billowed layer below. This view is over Belgium at about 10,000 m.

**11.3.2** The shadow of the contrail falls on a thin layer of altocumulus, below which there is small cumulus formed by ordinary midday convection. Only the shadows give an idea of the relative position in a still picture.

**11.3.3** This is the shadow of a contrail observed from the aircraft making it on a mixture of ice and water cloud. The ice cloud was dark when seen from above (see 4.2.6); while where the water cloud was not shaded by the cirrus it was bright white, and a glory could be seen on it around the aircraft shadow at the head of the contrail. This is a view from about tropopause level over the Boothia Peninsular (70°N, 90°W) at about 6·30 (local) in July. Similar shadows are common to see from jet passenger aircraft.

**11.3.4** The older higher contrail is less white because it is more diffuse and has begun to be spread by wind shear (see 11.4). In the lower one the vortices were still visible, and it was less than a minute old. Even so it has become distorted in what must be irregular air motion. Since the top is distorted with the bottom of it, the distortion is not due to the blob-forming motion. The shadow of the upper on the lower contrail is in late afternoon January sunshine over Kensington.

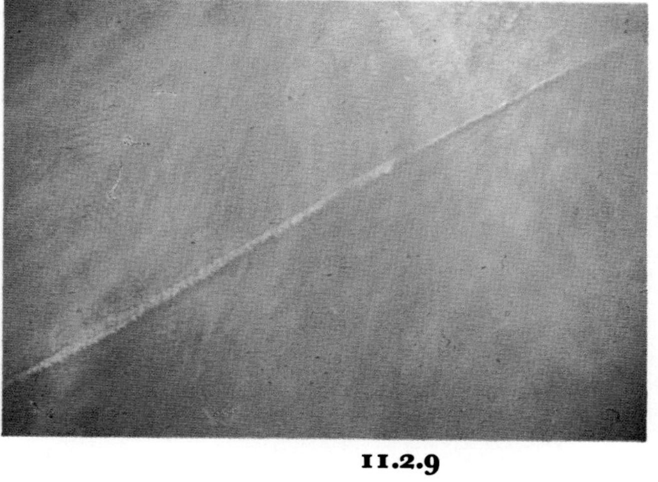

11.2.9

11.3.1

11.3.2

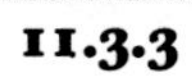

11.3.3

11.3.4

## 11.4 Glaciation and Distortion

Except for trails in wingtip vortex cores most contrails are formed at temperatures below −40°C (see 11.2.iv) and therefore the particles freeze almost immediately. Mock suns are often very bright in contrails, and parts of circumzenithal arcs have been observed in them.

Once glaciated the trails may last for many hours, and may be spread out by wind shear into almost horizontal sheets. The most extensive sheets are formed from trails which originally produced an extensive vertical curtain of cloud or very elongated blobs.

The trails persist in air that is saturated for ice and not for water, and such air is often found ahead of warm fronts. Often contrails can extend the forward edge of warm front cirrus by many tens of miles.

**11.4.1** These contrails are of various ages and are seen almost vertically overhead about three hours after sunrise in July at Wimbledon. The newest trail is the one from the top left corner: the oldest is the horizontal one, which has a delicate structure for which there is no obvious explanation.

**11.4.2** This trail was seen above the sunset over Bognor Regis in July. The blobs were extended over about a mile horizontally and in the shear layer billows can be seen.

**11.4.3** Old ice trails not subjected to much shear begin to acquire a fibrous structure suggesting more rapid fall of the particles along the centre of the trail. However contrails do not usually show the effect of the fallspeed of the particles because the particles are always very small and very multitudinous. This is because of the plentiful supply of nuclei and the rapid freezing of all the particles. Only if a very few froze would fallout be observed, and trails are not commonly formed at temperatures warm enough for this to happen.

**11.4.4** The aircraft making the distrail over Seattle in July in this unfrozen cloud evidently initiated glaciation so that a few particles froze, grew rapidly, and began to fall out. The ice cloud below the distrail is not a contrail and never showed blob formation.

**11.4.5** The upper of these two trails has an exceptionally well marked upper edge to the curtain of cloud, but not well marked blobs. This suggests that the cloud originated from two centrally placed exhausts. Some stretching by shear of the vertical curtain is occurring, and the lower, more distant trail has already been considerably extended.

**11.4.6** Contrails distorted by wind shear sometimes exhibit a screw-like structure. No satisfactory explanation has been proposed on the basis of a symmetrical aircraft wake, and this appearance may be the result of manoeuvres of an asymmetrical nature by the pilot. In this sunset picture, over Bayswater, low level very thin wave clouds are beginning to form at a haze top.

11.4.1

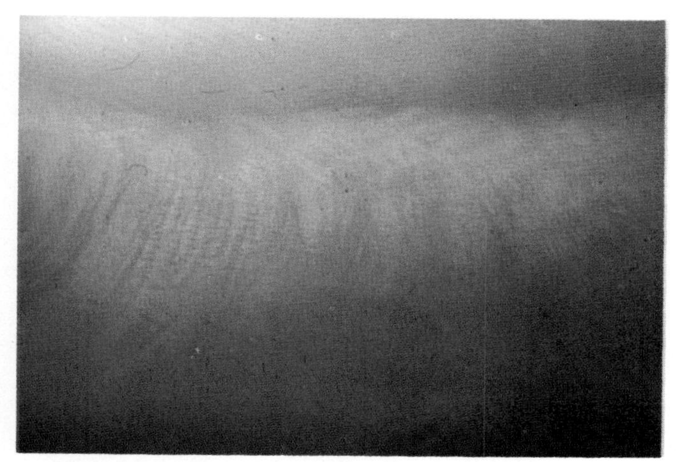

11.4.2

11.4.3

11.4.4

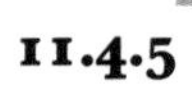

11.4.5

11.4.6

# 12 DROPLETS AND WINDBLOWN MATERIAL

Particles which are large enough to fall to the ground are not generally visible except when very plentiful. Thus rain in a shower can be seen, but particles of grit which might fall into your eye when you look up at the plume from a large chimney cannot be seen: the cloud of fine ash which forms the visible plume is composed of particles whose full speed is negligible and they follow the air movements, like smoke.

To capture small airborne particles a solid object must have small dimensions. Thus snow is deposited on the windward side of a tree trunk but fog particles are more effectively captured in a light wind on pine needles, spiders threads and other tiny objects.

## 12.1 Captured Ice

When the temperature is below freezing a twig or wire exposed in the wind captures supercooled cloud droplets which become attached to it by freezing. Large supercooled raindrops spread over the surface of the object before becoming fully frozen, and thereby form a layer of glassy ice. Very small supercooled cloud droplets freeze almost immediately on impact because the latent heat of freezing is more readily got rid of when the surface to volume ratio of the droplet is large. The accumulation of ice on the windward side of the object is then opaque and white because of the air spaces between the frozen droplets.

In a wind containing supercooled cloud droplets the air must be saturated with respect to water; if the temperature is more than about 15°C below freezing there may be a significant growth of the deposit by condensation of the vapour on to the ice accumulation. This will play a significant part in the form taken by the growth if the amount of liquid water captured and the size of the droplets is small, as would be the case just above the condensation level of the air forming the cloud.

These pictures were taken at the Milesovka Observatory (827 m.) in Bohemia.

**12.1.1** In the absence of wind, hoar frost grows by condensation on solid surfaces. The crystal form is often determined by the nature of the surface, and supercooled dew is virtually unknown. This example shows ice needles growing on a wire.

**12.1.2** The second most common form of crystal is plates, and in this example they have grown profusely on wire netting during a cold calm night.

**12.1.3** In a wind the growth is predominantly on the windward side of the object. This picture shows an accumulation of soft rime, in which direct condensation at a rather low temperature probably played an important part in determining the shape.

**12.1.4** The growth in this case differs from 12.1.3 probably because the crystal form is different on this solid surface. Like 12.1.3 the wind is light, the liquid water content of the cloud is small, and growth by direct condensation important. Deposits of this kind are very easy to remove by shaking the wire or brushing the crystals.

**12.1.5** In this case the rime was relatively hard, probably because of a greater amount of growth by capture of cloud droplets than in 12.1.3 and 4. This growth is bifurcated; it reaches a maximum extent (of about 20 cm.) not directly into the wind but about 30° on either side of that direction. The featherlike form suggests that even in this case growth by condensation plays a significant part.

**12.1.6** This accumulation occurred over about two hours during a day of 'fog showers' in which cumulus clouds whose base was below the observatory passed over leaving a rime deposit. The layers are produced by separate "showers" and it appears that they contained varying amounts of pollution. The wind was blowing from the direction of industrial sources and was considerably stronger than that in which 12.1.3 and 4 were formed.

12.1.1 12.1.2 12.1.3

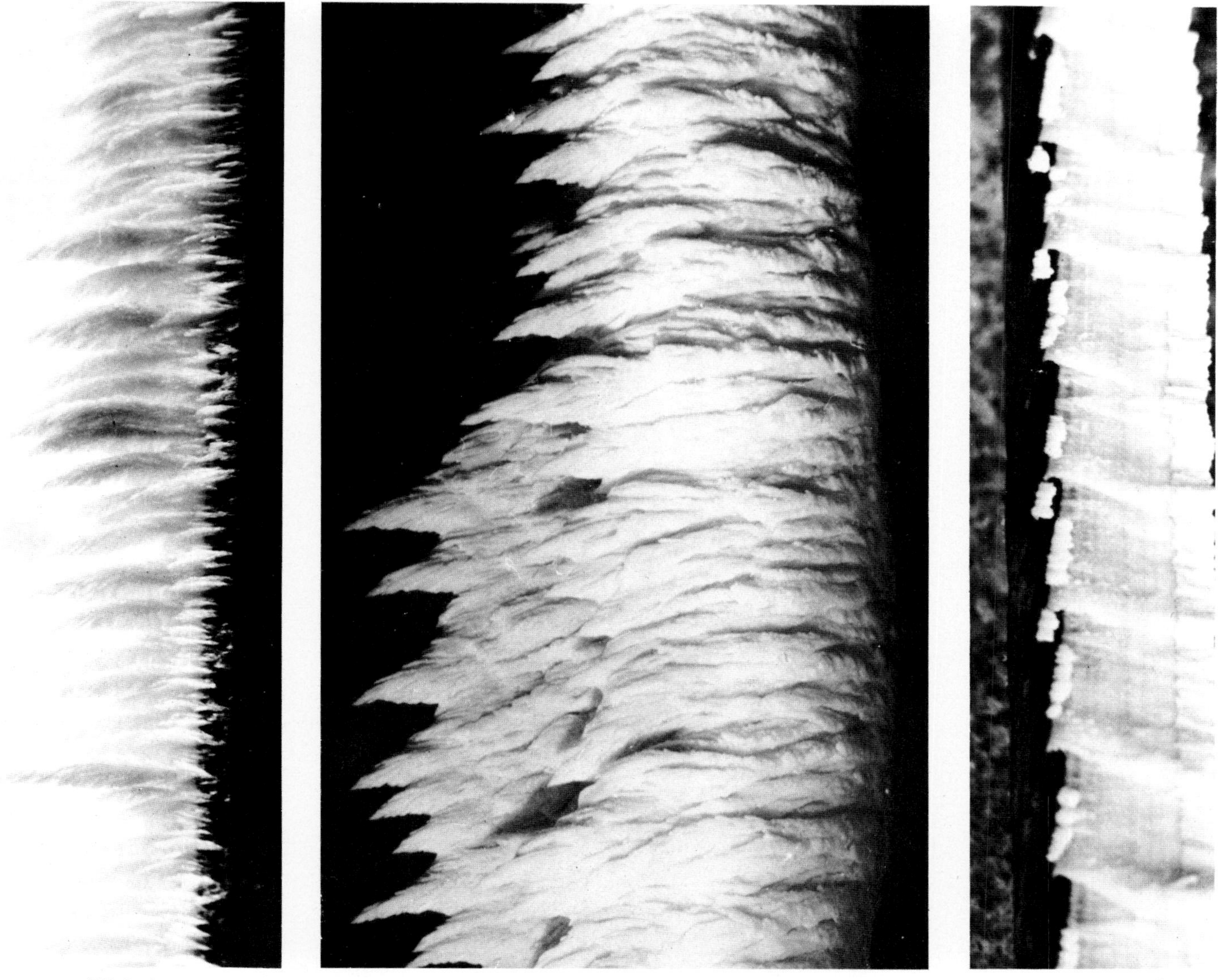

12.1.4 12.1.5 12.1.6

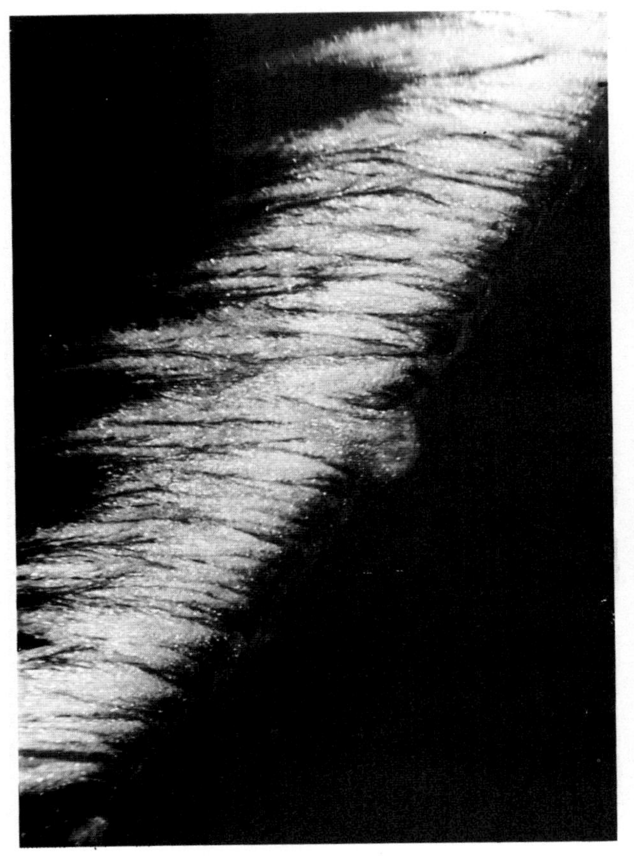

**12.1.7**

**12.1.8**

**12.1.7** Rime is glazed when the liquid water content of the cloud is high and the droplets spread over the capturing surface before freezing. At temperatures not much below freezing direct condensation of ice plays a negligible part in the growth.

The wind in this case is inclined to the wire, and it is likely that a growth initially more like 12.1.4 has become glazed over. Glazed deposits are often very strongly attached and ring when tapped.

**12.1.8** This glazed rime growth on grass blades (about 10 cm. high) under conditions similar to 12.1.7 shows no influence of condensation. The deposit widens as it grows, but is not bifurcated and therefore probably depends more on droplets which are larger (compared with the obstacle size) than those in 12.1.5.

**12.1.9** The accumulation of frozen rain on these twigs is about 1 cm. thick. There is no evidence of wind, and the supercooled raindrops spread, often dripping, over the surface before freezing.

**12.1.9**

## 12.2 Water Droplets

12.2.1

12.2.2

**12.2.1** This spider's web was observed after the clearance of the steaming fog in 9.4.1. The particles were captured as a result of the light wind.

The blades of grass did not capture a significant amount of the cloud droplets: the spider's web is very effective because of its small dimensions.

**12.2.2** When the air becomes much colder than the roots of grass the moisture fed to the leaves by the roots is exuded as guttation and forms large drops on the tips of the blades. This is typical of a summer morning in Wales.

12.2.3

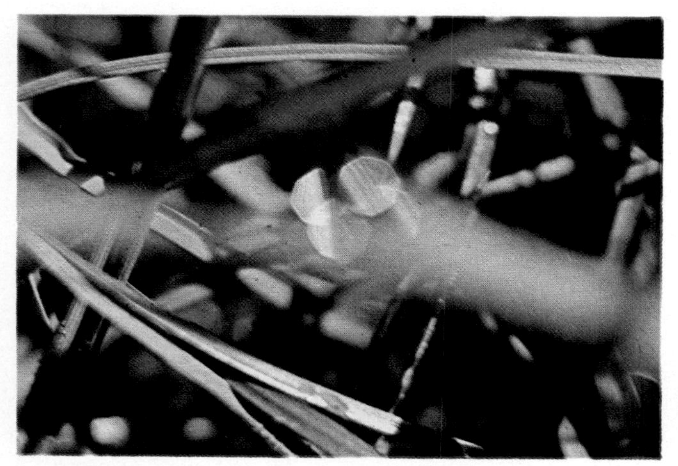

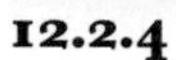

12.2.4

**12.2.3** Dew, which is water condensed on to a surface which has become cooled by radiation, usually at night, forms as a multitude of small droplets along the blade. There is guttation on the blade tips.

**12.2.4** When photographed in sunshine out of focus the four nearly spherical drops on the main blade of grass (out of focus) show the rainbow spectrum spread out on the film in the camera. The two droplets on the left show the main bow colours and several supernumeraries, those on the right show the secondary bow, in which the colours are reversed. (See 5.2.2, 13.2.5 and 13.2.i.) The fact that these colours are in nearly straight bands shows that the light refracted in the central plane of the droplet produces the most concentrated beam.

### 12.3 Windblown Material

**12.3.1** The Harmattan is a hot wind which blows from the desert and is accompanied by a fog of minute particles. In the case illustrated, at Karachi, the particles are of a mica-like substance blown up from the Baluchistan desert. This kind of dense haze which makes the sun yellow, is often observed over the Red Sea, and is also observed frequently on the southern borders of the western Sahara. The particles are slowly deposited during several days when the wind drops. Many clays are composed of flat particles and were originally formed by windborne sorting of particles. Dunes contain much larger particles which are carried only short distances above the ground.

**12.3.3** At the coast where there is a large concentration of persistent breakers the spray particles are so copious that they may be carried two or three miles inland, before being finally deposited, by a strong turbulent wind. This view is at the well-named town of Saltburn: the sky is yellow with salt haze. A wind of this kind in early summer can cause the scorching of growing vegetation, by the salt, especially of plants like beans which have tender leaves. In winter snow on the ground melts more quickly at the coast because of the salt.

**12.3.2** A blizzard is a storm of snow blown up from the ground. The most dangerous ones occur very suddenly when a strong wind aloft suddenly penetrates to ground level by disturbing a 'pool' of stagnant cold air. In such cases the blizzard can suddenly replace calm bright sunshine. Sometimes wave clouds are a sign that blizzards are likely because they are often accompanied by local gales on the lee side of mountains: the wave motion disturbs the calm air at the lowest level. This picture was taken in Greenland.

**12.3.4** The particles of spray blown off the wave tops by a strong wind are not an important source of airborne salt particles because they are too large for the water to evaporate before they fall back into the sea. But foam produced by breaking waves produces very fine particles by the rupture of the thin films which surround the larger bubbles. Also when air bubbles break through the water surface the surface energy released throws up one or two very small droplets. These minute salt water droplets quickly become damp salt particles and are carried into the air by eddies. On this occasion the very sharp horizon shows that the air is relatively dry and free from salt particles.

# 13 OPTICAL PHENOMENA

The atmosphere is full of subtle effects of illumination and of colouring by scattering, many of which cannot be properly depicted in colour printing because they are best seen in changing lights and with movement of the eye. Some involve physiological effects which cannot be photographed.

Clouds are neutral, but acquire a variety of colours from a variety of sources of illumination. The clouds may be bluer or redder at sunset according to whether the sun or the blue sky illuminates them; sometimes they are green by illumination from the countryside. The sky may be green, and the light at emergence from a hailstorm may be likewise tinted.

In this chapter we are mainly concerned with the separation of colours and the concentration of light by cloud particles.

### 13.1 Surround of Observer's Shadow

There are two effects which brighten the ground around the observer's shadow. First, no other shadows are seen there so that the ground is uniformly bright (see 8.3.1). Secondly water droplets which may be supported on grass focus the sunlight as a bright spot on the grass (like the sphere of a Campbell-Stokes sunshine recorder) and this bright spot acts as a source whose rays are focused as an almost parallel beam back towards the sun, i.e. towards the observer, who sees his own shadow carrying a 'halo' while other people's do not.

**13.1.1** The grass halo (Heiligenschein) is best seen when there has been a heavy dew, or better still a very fine drizzle which produces larger droplets on the grass. Guttation does not help because (see 12.2.3) the sunlight on it is not focused on to a surface. Grass which has whiskery blades is more effective because the bright spot is focused at a small distance from the droplet which is on the blade if the droplet is held at this distance.

12.3.1

12.3.2

12.3.3

12.3.4

13.1.1

## 13.2 Water Droplet Arcs

**13.2.1** The observer's shadow is surrounded by coloured rings, known as the glory, when it falls on a cloud. The rings are blurred if there is a wide range of drop sizes present. In this case the cloud is very thin with a small drop size range and four or more rings can be seen. In theory rings might be seen all the way out as far as the cloud bow in a cloud of uniform drop size.

**13.2.3** The angular distance of a particular colour from the sun depends on the dropsize and whether it is the first or an outer ring. Close to the sun the colours tend to be in rings around the sun, as here, but at greater distances they tend to show contours of constant dropsize in the cloud. (See 5.4.5.)

To see iridescence it is helpful to obscure the sun and use dark glasses and give a very short exposure in photography. This picture shows a wave cloud on a summer afternoon over London.

Iridescence is commonly seen up to 25°, and sometimes up to 40°, from the sun. (See also 11.1.4.)

**13.2.5** In rainbows the common, primary, arc is produced by one internal reflection of the light rays. The returning rays are confined within an angle of about 41° from the antisolar point, and on the boundary there is a concentration of intensity, at an angle whose exact value depends on the colour, which produces the rainbow. There is very little light reflected back between this arc and the secondary at 52° (which is produced by two internal reflections) where the colours are reversed. (See 5.2.2 and 13.2.i).

Inside the primary, supernumerary bows are often seen, and these are most intense when there is a small range of drop sizes.

**13.2.6** At sunrise the sky is orange. This sunset picture, taken at Milford Haven, shows a rainbow against a cloud. The cloud shadows produce anti-crepuscular rays which converge on to the antisolar point just below the horizon. Such rays and bows are common at sunrise over the tropical oceans where warm rain is formed in among evaporating clouds above cloud base.

**13.2.7** This cloud bow seen through a 28 mm. lens was photographed on fog on the slopes of the Pic du Midi from a cable car. The green grass below colours the scene. The shadow of the car and a faint glory are seen at the centre of the arc.

**13.2.2** The corona is a similar phenomenon, produced by small cloud droplets around the sun. In this case patches of cloud were passing over the mountain top and the sky, which was blue, with the sun behind the pillar, became intensely bright. In this short exposure the blue sky on the left appears black. The mountain is Cader Idris.

As in the glory more rings are visible when the range of drop sizes is smaller.

At greater angular distances the phenomenon is called iridescence (see 5.4.5).

**13.2.4** It is important to shade the camera lens from direct sunlight to avoid multiple images. Perhaps more important is to avoid letting sunshine reflected from a shiny part of the lens mount enter the lens. This picture was taken of crepuscular rays in smoke in the shadow of trees, but they do not appear because they are too faint. Instead we see rays from part of the lens mount. The colours are caused by diffraction in the scratching on the metal. Similar colours can sometimes be seen on ones own eye lashes in sunshine due to thin films covering them. What is seen in the camera viewfinder is not necessarily what the film will see because, even in a single lens reflex camera, a pentaprism, its mirror, and mountings can produce false rays and images.

In this case the rays emanate from a point to the right of the sun's position shown by the tree shadows, and, of course, the smoke patch was not cone shaped, and could not have produced the colours here recorded. Films may also produce false rings. (See 13.3.21.)

**13.2.8** A cloud bow is formed in about the same position relative to the observer as a rainbow, but it is less brightly coloured. This one (seen with a 35 mm. lens) is an exceptionally clear case, seen on the same occasion as 13.2.1 which was photographed with a narrower angle lens (50 mm.). This cloud, over Essex, was a very thin layer giving much less than average background illumination. Because the cloud bow is layed out on a horizontal sheet of cloud it is not readily recognised as a circular arc. On the cloud surface it forms a hyperbola when the sun is below 84° elevation, which is the most common case. The bow is highly polarised, and supernumeraries inside the main bow are common. Secondary cloud bows are very rare.

13.2.1

13.2.2

13.2.3

13.2.4

13.2.5

13.2.6

13.2.7

13.2.8

**13.2.i** Paths of rays through spherical water drops forming the primary rainbow (Rp) at 42° and the secondary (Rs) at 51° from the antisolar point.

**13.2.10** A fog bow (Ulloa's ring) is a cloud bow. This one which has a bright first supernumerary was seen near Asmara in cloud similar to that shown in 9.2.3 but in the evening.

Other bows can be formed by reflection of sunlight in a calm lake.

### 13.3 Ice Crystal Arcs

**13.3.1** The very common 22° halo (13.3.iii) is produced by refraction through randomly oriented prisms. Most rays will be deviated through a greater angle than this so that the sky is brighter just outside the halo than just inside. The inside of the arc is tinged with red, which is less deviated. The random orientation is probably caused by departures from simple regular hexagonal plate form.

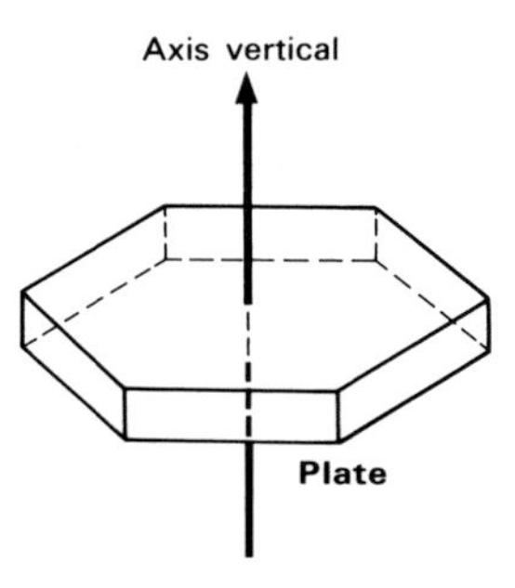


**13.3.i**

**13.3.i-vii** The ice crystals which produce the most common arcs are hexagonal prisms. We call them needles (ii) when their axis is much longer than their diameter, and plates (i) when it is much shorter. Most of the refractions possible in needles are possible in plates which have plane upper and lower faces, because, by total internal reflection at these faces (v) the plate prevents escape of most of the light from them. It depends on the angles whether plates can produce arcs as bright as needles.

The least deviation possible on passage through faces 60° apart is about 22°. Plates usually have their axes vertical and needles have theirs horizontal. The

**13.2.9** This fog bow with one supernumerary inside, and a secondary faintly visible outside, was photographed at about 10 am from the top deck of the Queen Elizabeth 2 (which is a superb observing point). The fish-eye lens used (180°) distorts the shape of the ship, whose smoke passes in front of the fog making the bow.

**13.3.2** The halo at the top is well coloured. Below is a circumhorizontal arc which is formed by refraction in horizontal rectangular edges of needles or prisms. The colours are clear when these all have vertical axes, but would be blurred, except for the upper edge, by abundance of horizontal axes producing the 46° halo's lower tangent arc. The sun must be more than 58° above the horizon or the rays will be totally internally reflected at the second (horizontal) face (13.3.iv). The arc is parallel to the horizontal just above the framework tower and has the red uppermost (see 13.3.17 and 18). It is not common, nor is it bright compared with the rest of the sky, and the colours have not survived reproduction in this case, photographed at Brookhaven, L.I.

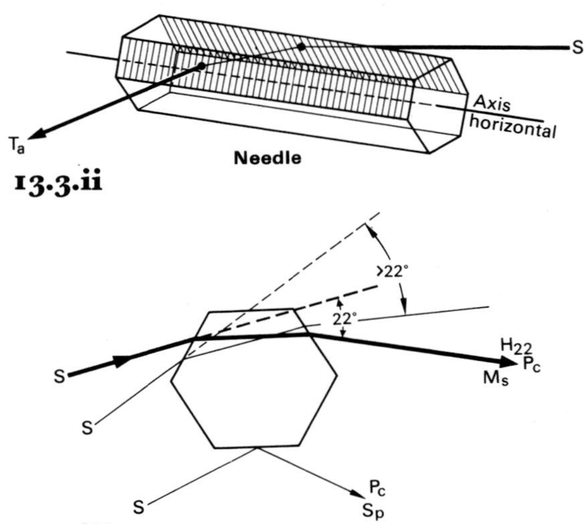


**13.3.ii**

**13.3.iii**

paths of rays are sketched for the phenomena illustrated. H22 = 22° halo &c. Ta = Tangent arc (upper 22°) and circumscribed halo. Ms = Mock sun, Pc = Parhelic circle, Ch = circumhorizontal arc, Cz = Circumzenithal arc, L = Lowitz arc, Pac = Paranthelic circle, Sp = Sun pillar, Ss = Subsun, S = sun.

In (iv) and (v) the axes of the prisms are vertical. The elongated prisms in (vi) rotate around their horizontal axes to produce the Lowitz arc (13.3.3) and the prisms in (vii) are randomly oriented to produce the 8° halo (13.3.20) and have either pyramidal extensions or hollow pyramidal ends: both of these are very rare.

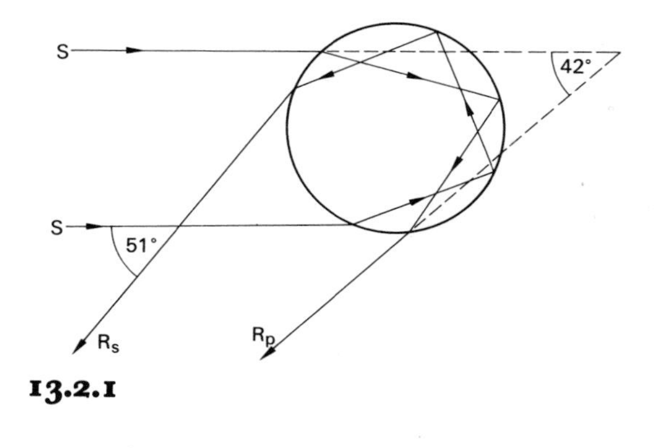


13.2.1

13.2.9

13.2.10

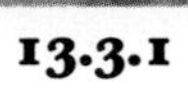

13.3.1

13.3.2

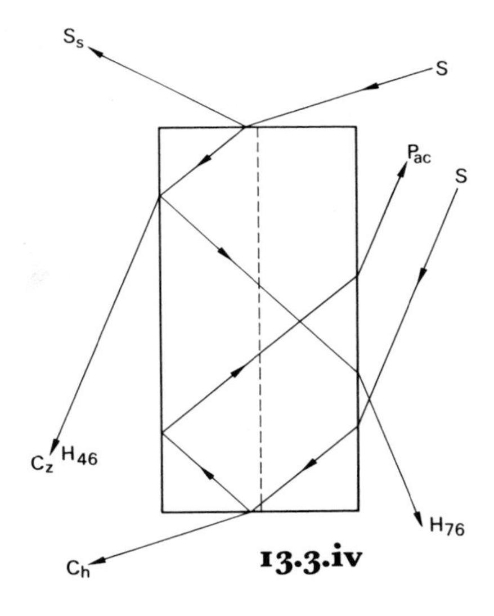


13.3.iv

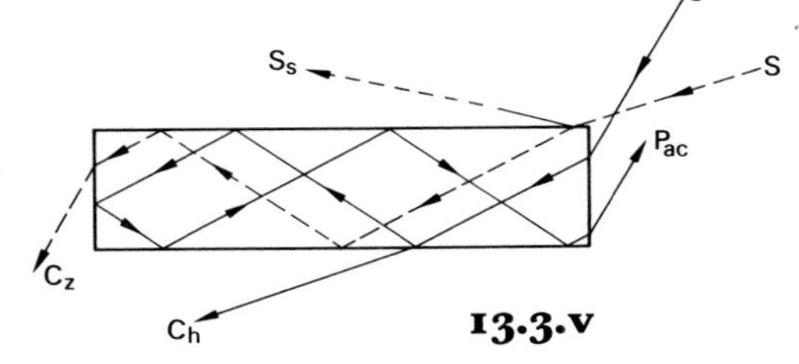


13.3.v

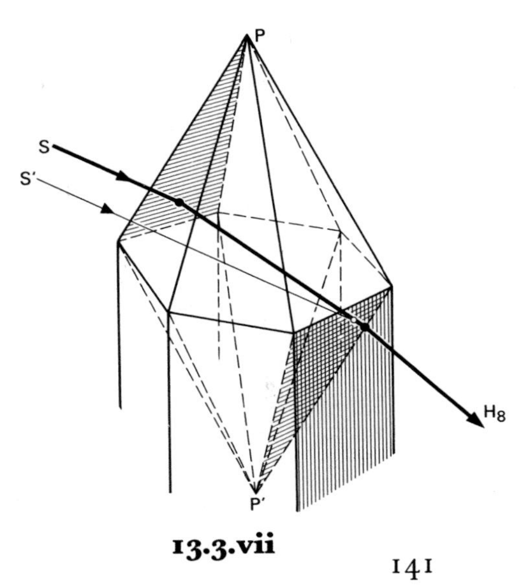


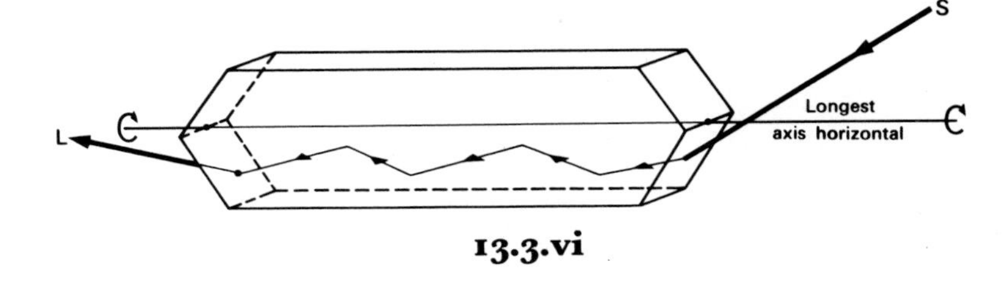


13.3.vi

13.3.vii

**13.3.3** The 22° halo is seen faintly inside the circumscribed halo which joins it tangentially at the uppermost and lowest points. There is a predominance of needles with their axes horizontal in this case, and the minimum angle of deviation through faces at 60° exceeds 22° when the crystal is not in the vertical plane through the sun and the observer and the plane of refraction is not perpendicular to the crystal edges (see also 13.3.13, 14, and 15). There is also a white parhelic circle (see 13.3.4, 7, and 8) in the right half of the picture.

The sharply curved arc which is tangent to the 22° halo on the right and crosses the circumscribing halo at about 4 o'clock is the Lowitz arc. The only plausible explanation is that due to R.A.R. Tricker, according to which the arc is produced by plates which are elongated, not regular hexagons (13.2.vi). These would rotate about their longest axis which would be horizontal. The rays selected for refraction in this way accord with the observations. Such crystals must be very rare because all the usual mechanisms tend to produce regular hexagons: the arc is extremely rare as would thus be expected.

The contrail across the sun has been spread by wind shear (see 11.4). It lies above the halo-producing cloud: this was shown when it was seen to cast a shadow on the cloud as it drifted across the sun. This shadow can be just seen on the right of the upper and lower extremities of the trail.

**13.3.4** The white cloud on the left is supercooled, but where it has become glaciated it is more transparent to the blue skylight because the number of cloud particles is much less. The spread of the glaciation into the water cloud along narrow 'rivers' can be seen. There is a mock sun in the lower part of the ice cloud. This is formed by refraction through faces at 60° in prisms with vertical axes. The sun is to the right, and so is the red. The other colours are smudged over and become white on the left of the mock sun.

**13.3.5** There are very many possible arcs and bright spots of light. This "halo" surrounds the antisolar point (observer's shadow) and was observed in Antarctica where optical phenomena abound. There is no simple explanation, and it may be due to some crystal form which is common only at rather low temperatures: it was −50°C nearby on this occasion. A single reflection in a hexagonal crystal would produce a 38° antisolar halo. This one had an angular radius of about 25°. In appearance it is very like a cloud bow (13.2.6-8) and it could readily be explained as such in spite of the small angular radius if there were very small drops of rather uniform size. Even if the fog had only just formed at the low temperature observed it would normally have frozen immediately. If the air were very stable the measured temperature could refer to different air.

**13.3.6 a, b** These two pictures taken on the same occasion show the effect of having only crystals whose axes are vertical or horizontal present. The circumscribed halo (upper tangent arc) lies above the sun whose altitude is about 40°, and the mock sun is a bright patch at the end of the parhelic circle. There is a contrail in *a* which has just been formed: in *b* it has moved to the concave side of the parhelic circle. The sky is very bright, generally, so that the arcs are usually conspicuous mainly because of the faint red on the inside of the arc and the pure whiteness of the horizontal circle by comparison with the general blues and greys of the rest of the sky.

The parhelic circle, which is always less bright inside the mock sun, because it is produced only by external reflection there, may be absent completely because of the absence of crystals there or because they are mostly thin plates. Many of these arcs are the first evidence that any ice particles are there at all (e.g. see 13.3.8 and 18). Parhelic circles may extend nearly all the way round the sky, and sometimes the parts opposite the sun may be the only parts visible.

13.3.3

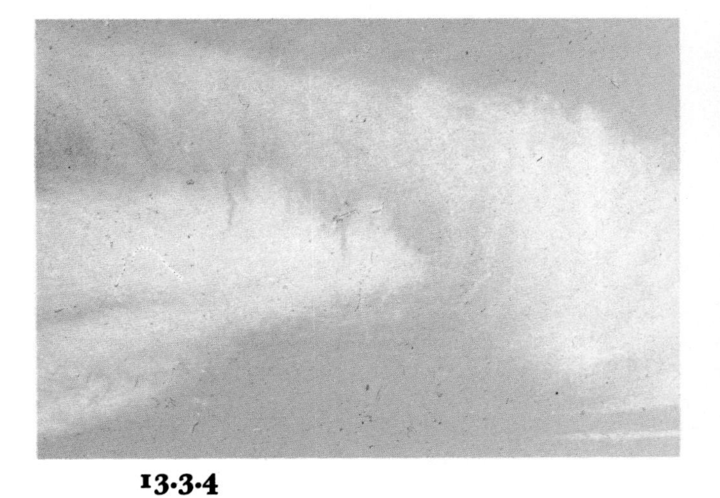

13.3.4

13.3.5

13.3.6a

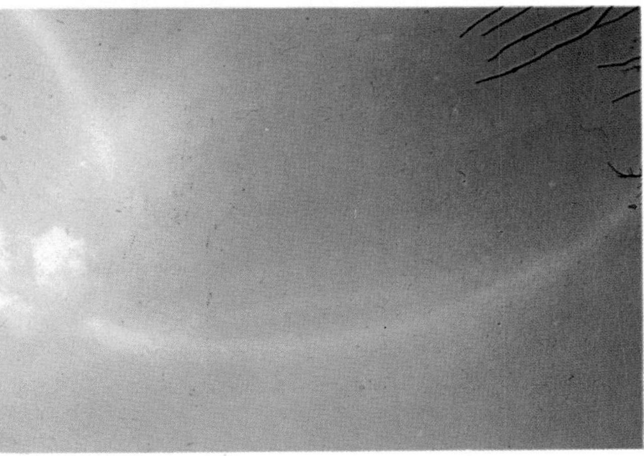

13.3.6b

**13.3.7** In snow and ice covered territory halo phenomena are common because most clouds are frozen. In this picture taken with a 200° lens in Antarctica there are three concentric haloes at 22°, 46° and about 76°. The 46° halo is the angle of minimum deviation of a ray refracted through faces of a crystal at 90° to each other. (See 13.3.iv.) The explanation of the 76° halo is less certain (see 13.3.iv); the halo itself is rare. If the explanation given is correct both the 46° and 76° haloes could be formed with horizontal ice needles oriented freely in the horizontal plane and rotating about their axes. The 76° halo is a maximum intensity arc, not a minimum deviation arc, according to the present explanation, and it is necessarily rather faint.

Through the sun and parallel to the horizon lies the parhelic circle, formed by reflection in vertical crystal faces or refraction through two vertical crystal faces. The colours are only separated at the mock sun which is the extremity of the arc for refracted rays, for elsewhere the colours overlap (see 13.3.6). The mock suns of 22° lie just outside the 22° halo; the halo there being formed by crystals whose axes are perpendicular to the line from the observer to the sun.

At the top of the 22° halo is the tangent arc (or circumscribed halo, see 13.3.14) which is formed by prisms with horizontal axes. At low sun altitudes this tangent arc is concave upwards where it touches the halo and then curves downwards, becoming very faint beyond about 35° azimuth from the sun. It is visible as a bright spot at the lowest point of the halo.

There is also a 'tangent arc' to the 46° halo, but this is probably a circumzenithal arc formed in crystals with vertical axes. The 46° tangent arc and parhelia would have to be formed, like the Parry arcs, in horizontal needles with vertical and horizontal 90° edges respectively. But these arcs and the upper tangent arc could not be distinguished from the circumzenithal arc. The presence of the 22° parhelia proves the presence of crystals with vertical axes, and these would produce the circumzenithal arc. (See 13.3.18.)

Since these tangent arcs are more highly coloured than the haloes, the crystals having one less degree of freedom, they may be confused with other arcs when the haloes are not present. Among these are the very rare Parry arcs which are formed by needles with horizontal axes with the further restriction that they have two faces vertical or two faces horizontal, which are slightly more stable positions than positions in between. They lie outside the halo, or tangent to it when the sun's elevation is 30° ±11° in one case or 60° ±11° in the other. The tangent arc to the 46° halo is in almost the same position as the circumzenithal arc which is much more common than that halo (see 13.3.18 and 19). When the halo-producing crystals are present, the part close to the halo is intensified in brightness but the colours are somewhat smudged. Other faint arcs have been identified by Tricker in this picture, but it is difficult to print it with the contrast to show them all together.

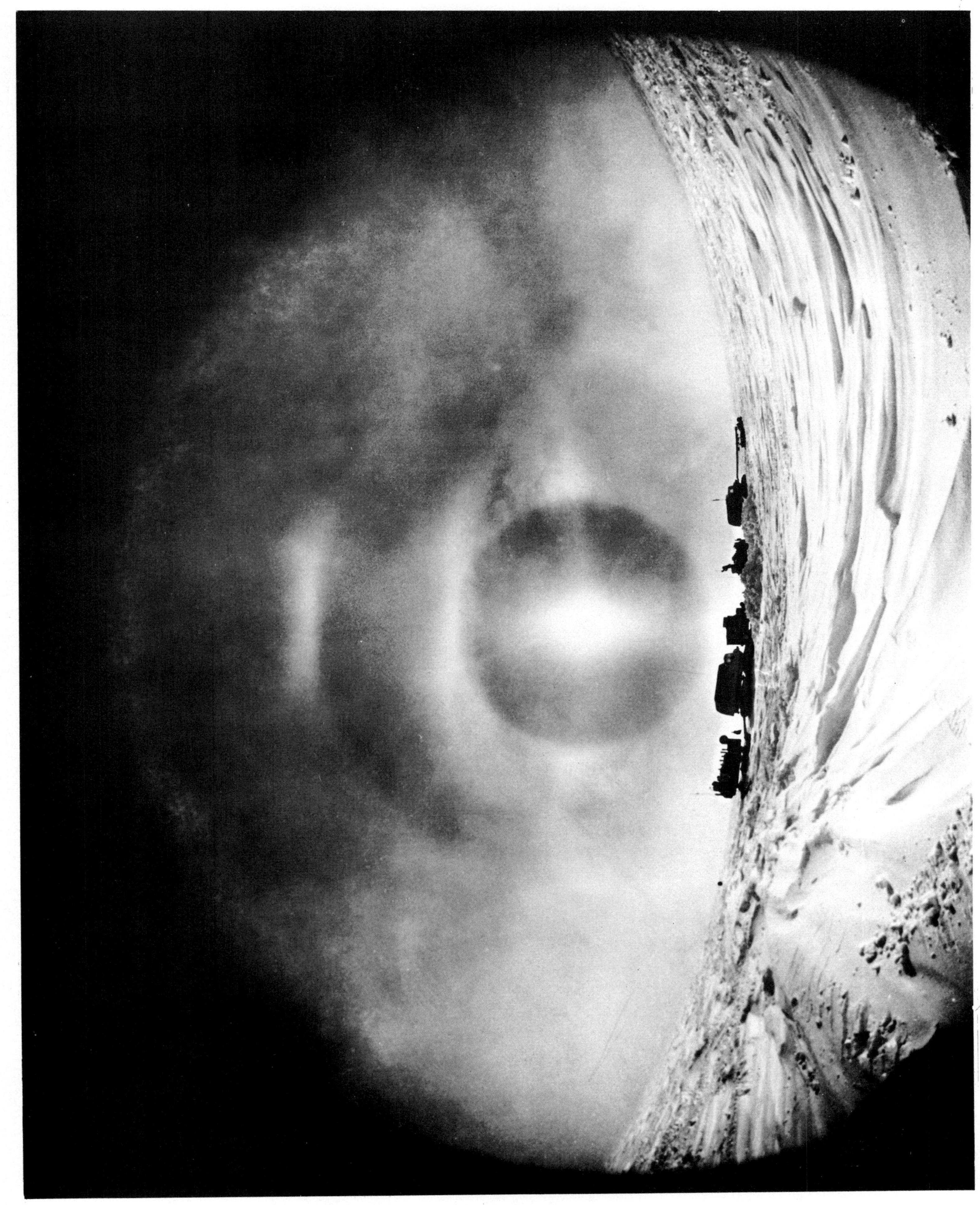

13·3·7

**13.3.8** The camera was tilted on this occasion in order to get both mock suns into the picture. The sun was hidden behind a theodolite to avoid glare. In this case the inner parhelic circle and halo are much brighter which shows the presence of randomly and vertically oriented crystals. Place—Antarctica.

**13.3.9** This arc is evidently very rare, no other report of it being known. It is not easy to see because it is necessary to be above an ice cloud, and from an aircraft it is not easy to look steeply downwards. It is a horizontal white arc passing through the observer's shadow and is therefore a paranthelic circle. In theory it probably has bright spots (paranthelia) and could be observed passing through a subsun (13.3.10, 11, 12) but this is not known to have been reported.

The part of the arc seen here, near Rome, passes through the aircraft's shadow (bottom left), and was formed either by the emergence of rays from the face of entrance after reflection at the lower and opposite faces of a vertical axis crystal (see 13.3.iv and v) or by external reflection on vertical needles with plates at their bases. This arc would be expected to occur when circumzenithal arcs are visible, and at other times when the sun's altitude is too great for that arc to be formed. Therefore it should be looked for often.

**13.3.10** Subsuns are quite common. They are formed by reflection in the horizontal upper surfaces of plates with vertical axes. Because the plates' axes do wobble somewhat around the vertical the spot of light is spread, and an equal wobble in all directions produces more displacement of the image in the plane of the observer and the sun than at right angles to it. The subsun is therefore extended in the vertical plane, and to a greater extent the lower the sun. On the right is a mock subsun, which is less common. It may be formed either by light from a subsun passing through a second crystal or by internal reflection in the horizontal base of a crystal forming an ordinary mock sun.

**13.3.11** The ring around this subsun has not been satisfactorily explained. It has the same elongation as the subsun, and did not change form when viewed through different parts of the window. It was seen, near Des Moines, Iowa, on the same occasion as 13.3.10, and the ring was present only part of the time. No colour separation was visible, so that it was not a diffraction phenomenon. It may have been due to a small hollowing out of the upper surfaces of some of the crystals.

The bright spots beyond it are specular reflections from water on the ground. Such a ring around a submoon could well stimulate the imagination of the UFO seeker.

**13.3.12** To the left of the centre of this picture is a well coloured mock subsun, seen beneath a layer of cloud over the partly ice-covered Beaufort Sea. The clouds there are unusual in that many of the very low clouds are frozen while those above are supercooled. This mock sun may be rather bright and well coloured because it is formed in the usual way from an image of the sun in a patch of calm unfrozen sea illuminated through a gap in the clouds. The background is very much darker than is usual for a mock sun because of the cloud shadows, and it appeared almost as bright as a rainbow. In stereo the mock sun appears to be at a great distance and its luminosity is enhanced.

**13.3.13** From the level of ice clouds haloes, mock suns, and sun pillars are commonly seen when the sun is low. They are often very bright because of darker bluer backgrounds and a greater proportion of blue light. The upper tangent arc becomes very concentrated at the top of the 22° halo when the sun is low (see 13.3.15) so that the brightening at that point seen here is very common. The sun pillar is faint in this case. There is a 'subsun' on the aircraft wing.

**13.3.14** Sometimes, when only needles with horizontal axes are present, the circumscribing halo is bright. When the sun is high in the sky (in this case it was about 50° above the horizon) the halo appears elliptical. The lower part which is not quite symmetrical is less commonly visible, and a display such as 13.3.3 is rare. This part of the arc is, however, very common, bright, and well coloured with red on the side towards the sun.

13.3.8

13.3.9

13.3.10

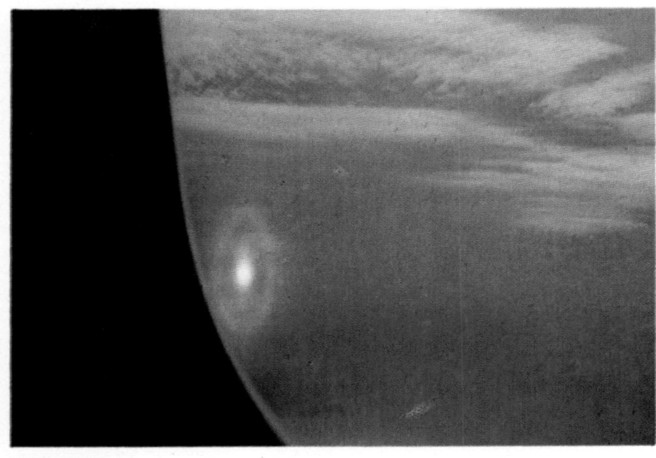

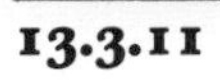
13.3.11

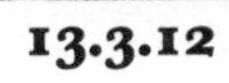
13.3.12

13.3.13

13.3.14

**13.3.15** When needles predominate the sun pillar and upper tangent arc are to be expected. With the sun overhead the arc coincides with the 22° halo but at lower altitudes it gradually becomes further separated until when the sun is on the horizon it is concave upwards. The oscillation of the needles about their equilibrium horizontal position spreads the bright spot at the junction of the pillar and the arc. This example was photographed in Jersey. The pillar does not have sharp edges but represents the direction of the most efficient specular reflection.

**13.3.18** A circumzenithal arc is formed above the sun, with the zenith at its centre, by refraction in the 90° edges of crystals with vertical axes. Because the crystals have only one degree of freedom this is a bright arc with all the colours present. The red predominates in most haloes because they minimise deviation phenomena and the other colours are smudged. The blue is difficult to identify in a circumzenithal arc because there is usually blue sky beyond, but the green is usually very clear. This example, from the Canadian Arctic, shows the presence of crystals in an apparently clear sky.

**13.3.20** Like many of the theoretically possible arcs the 8° halo is very rare, probably because the crystal form producing it is rare. Because of the small angle of deviation the colours are not perceptibly separated. It is difficult to see because of its nearness to the sun. The only other occasion on which it is known to have occurred was at a rather warm temperature, for the arc was photographed in freezing steaming fog over Lake Geneva. The occasion depicted here was over a snow surface in Canada and the halo may not have been caused by the high clouds at all, but by low level ice crystals. The rays are refracted through the face of a pyramid, either internal or external, and the opposing face of the prism (see 13.3.vii).

**13.3.21** This is a typical mock sun (parhelion, sun dog) which is a common phenomenon (and a possible form of UFO) with the brightest part of the parhelic circle also present (see 13.3.6).

The red ring around the sun is false. It is formed in the film, but the explanation is not known. In all colour films in which this ring has been found to occur it is red or reddish brown. It is best defined when the sun's image is bright and small. Thus it occurs with lenses of focal length less than 30 mm. rather readily, but not with

**13.3.16** Crystals abound in the air over loose lying snow, partly because they are carried up in the wind and take a long time to settle, but also because the air is continually being cooled by radiation. Any incipient fog is immediately frozen at these low temperatures in Greenland (though not over warmer snow covered ground) and becomes a very sparse collection of crystals. Optical phenomena abound but have not on the whole, been systematically photographed and reported. This sun pillar being inside the 22° halo is formed by reflection from needles with horizontal axes.

**13.3.17** The circumhorizontal arc is formed only when the sun's zenith distance is less than 32°. At this angle the arc is on the horizon. It is best when the sun is 22° from the zenith, and for about 6° on either side of this the arc is about 46° below the sun. It is therefore often difficult to see because of haze (eg 13.3.2). This excellent example was observed on Mount Olympus (2,424m) near Seattle.

**13.3.19** Circumzenithal arcs are formed with their nearest point to the sun 46° or a little more above it. They cannot be formed when the sun is more than 32° above the horizon because the emergent ray (see 13.3.iv) would then be totally internally reflected. When the sun is near 22° the arc is tangent to the 46° halo and is a little above it when the sun is at other angles.

This arc was seen over London and the straight cloud is a contrail.

f greater than 100 mm. However, it has been obtained around a spotlight with a 135 mm. lens. The circle's radius is independent of f and the circle extends beyond the edge of the picture frame, and on to the next frame in cinefilm. Occasionally a second and third sharp circle, rather yellower than the first, have been found at exactly twice and three times the radius. The false circle is always complete and uniformly intense around its circumference, so that natural phenomena which are less uniform can be identified as such. The circle is centered on the bright spot of light wherever it is in the picture, and is not a false image produced by the lens.

There is no natural mechanism which could produce the red ring in the sky except as the edge of a brighter white area, like the haloes. The ring can be avoided by shading the sun from the lens or letting it be blurred over a larger area by thin cloud.

13.3.15

13.3.16

13.3.17

13.3.18

13.3.19

13.3.20

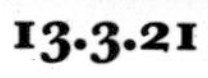

13.3.21

## 13.4 Illumination Effects

**13.4.1** After a large volcanic eruption an exceptional amount of dust is distributed in the stratosphere. But even this cannot be seen when the sun is shining on the troposphere, from which much more light is scattered. Just after sunset, when the lower layers of the stratosphere are still illuminated, probably by a rather reddish light that has passed through the troposphere further west, the glow of the dust shows a purple tint particularly in the darker part of the sky. Such glows may be observed for a month or two after the eruption, and are very variable as more or less dense clouds of dust are carried over by the wind.

**13.4.3** Noctilucent clouds are at a height of about 80 km., where the pressure is about 0·1 mb. ($10^{4}$ atmospheres). Like stratospheric dust, they are not visible by day, and are usually too tenuous to be visible if the light illuminating them has passed through the troposphere. There is therefore only a short time when they can be seen even if they are present all the time. This lasts for only about 10 minutes soon after sunset and before sunrise at the equator, but on midsummer nights in high latitudes they may be visible for up to two hours on either side of midnight. This example was seen in central Sweden in July. It is characteristically bright white, the cirrus appearing as black shadows against the glow of the midnight sun.

**13.4.5** This rain streak over London became visible when the sun shone at dusk under the distant edge of the clouds producing the rain. The sun is setting in the north west in mid June, and the view is towards the sun.

**13.4.2** When sunset is observed from a high mountain (in this case the Pic du Midi, 2,877 m.) the shadow of the earth may often be seen on the eastern sky. This is a 1 sec. exposure with a wide angle (24 mm.) lens at f3·5 showing the dark band across the sky which rose up from the eastern horizon: the sky to the west was clear. The lower layers are brighter because of the denser haze.

**13.4.4** Often noctilucent clouds show a billowed structure. This example was seen at Grande Prairie, Alberta. Observers have seen what appear possibly to be gravity waves travelling through patches of this sort of cloud, which suggests that they are a water condensation rather than a dust cloud; but the origin of such waves is unknown. Certainly they are not directly affected by steady mountain waves. Often billows of much shorter wavelength are seen. If a viewpoint can be found remote from other illumination noctilucent clouds can quite frequently be seen on summer nights, when the possible observing period is longest, as near to the equator as 50°. It is not known whether they are actually much less frequent in lower latitudes or simply less easy to observe.

**13.4.6** Interesting unusual illuminations of ordinary clouds often occur near the coast when the sunshine is reflected off the sea. In this case the base of a layer of anvil stratocumulus displays its mamma brilliantly on the west coast of Wales. The more common kind of mamma, namely negative thermals (see 3.6.6 and 8) are best seen at sunset, and some are so tenuous when all the smaller droplets have evaporated that they only become visible by oblique light from beneath as red or orange appendages on a dark cloud.

## 13.5 Refraction in Air

**13.5.1** Mirages are common over hot ground where oblique rays are totally internally reflected because of the decrease of refractive index towards the ground. They are less common in the sky. The conditions are most favourable when the ground is snow or ice covered and therefore remains cold while the upper air is warmed by subsidence in an anticyclone. This view was taken in the winter in the Erzegebirge with a stereo camera which showed the images in the sky to be at a great distance beyond the fence. (There is, however, some white glare in the foreground which spoils the picture.) The image is larger than in a typical superior mirage.

The images of distant objects were extended vertically by the total internal reflection in a warm upper layer in which the temperature gradient increased upwards (see fig 13.5.i, p152). More commonly the image is merely an inverted, but distorted, version of the distant object.

The great vertical extension of the images which is not particularly usual in a superior mirage is likely to be temporary, and gives a vision of great towers and castles in the distance. In the Straits of Messina the phenomenon has been called the Fata Morgana (Italian for Morgan the Fairy, a woman of Celtic legend, who dwelled in the castles). There, and in

13.4.1

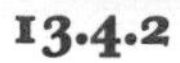

13.4.2

13.4.3

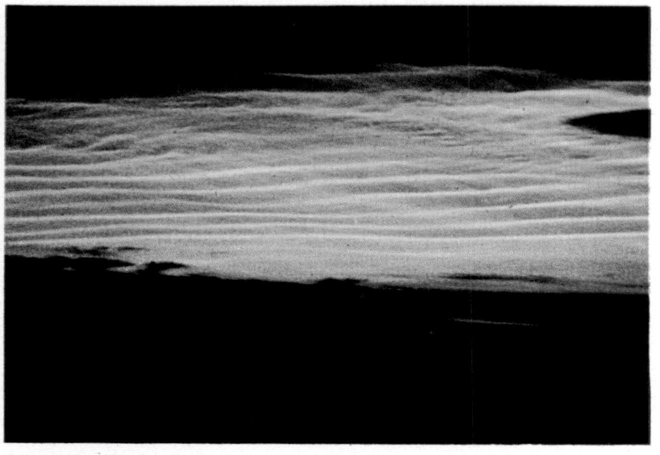

13.4.4

13.4.5

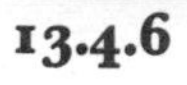

13.4.6

other parts of the Mediterranean, very strong inversions at low altitudes are common, and are occasionally low and strong enough to reflect the cliffs of a distant coastline in the sky.

Because of the great decrease in absolute humidity above the inversion the layer also reflects radar very effectively. The inversion and the sea form a radio duct along which coastlines far beyond the visible horizon can be seen by radar.

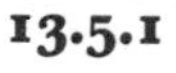

13.5.1

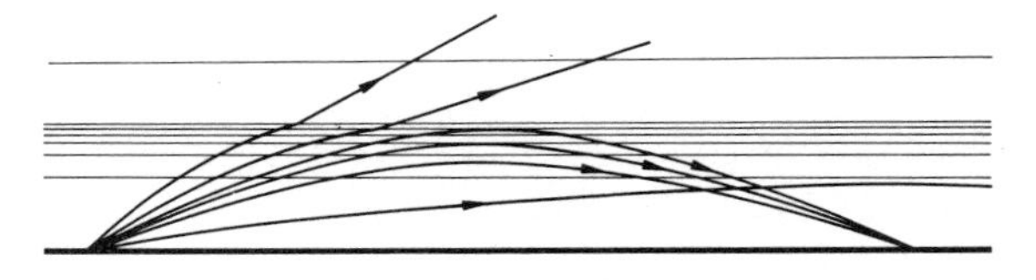

**13.5.1** The rays within a certain angle are focused at one point by the inversion; the image of that source of light is therefore extended vertically when viewed from the point to which they converge.

**13.5.2** The image of the sun becomes laminated by the variable refraction of the atmosphere. In *a* the sun is seen approaching the horizon. In *b* and *c* which had much longer exposures, the red image is separated from the other colours, so that the bottom edge of the sun on the horizon is reddened. At the same time the upper edge is tinged with green. The blue end of the spectrum is absent because of the long light path through moist or hazy air.

When watched through a telescope, and sometimes through binoculars these green edges can be seen most clearly just as the irregularities in the sun's image are disappearing as the sun sinks: they look like delicate coloured flames. In making these observations it is important to have the sun's image placed centrally in the eyepiece because colours produced by aberrations may be mistaken for the natural phenomenon. It should be remarked that the sun should only be looked at in this way when it is very low and not very bright, otherwise serious damage to the retina may be caused. The safest method is to attach a single lens reflex camera to the telescope and observe the image on the ground glass, and this is the obvious procedure when photographs are being taken. This is a view through a Questar on the west coast of Wales.

On occasions when the phenomenon is specially sharp and clear, the upper limb of the green component of the sun's light sets after the red by perhaps half a second. In such circumstances a green flash may be observed. Anecdotes abound about the green flash: it has been seen from the air and from mountain tops as the sun sets behind a bank of cloud. It is claimed that a blue flash has been observed in the clear air at the Pic du Midi as Venus descended behind a distant mountain: and further that, because its path of descent lay along a jagged mountain edge, seven successive blue flashes were observed on one occasion.

Several successive green flashes have been claimed for one sunset by observers in a boat moving up and down in heavy swell, but in assessing the validity of these claims it must be realised that the eye momentarily sees a green colour whenever a red light is switched off, indeed a red light flashing with a frequency of about 4 flashes per second can be made to appear green if the ratio of the flash length to the interval between them is adjusted to be about 1:8. Therefore these observers should have noted a green flash as they ascended each time from the wave trough as well as when they descended if it had been a genuine atmospheric phenomenon.

Physiological green flashes cannot be observed at sunrise, and so that is the time to make an uncontrovertible observation; but it is not a simple matter to observe the first moment of sunrise through a very narrow angle telescope on a sharp, clear horizon.

## 13.6 Polarisation

**13.6.1** The blue sky is highly polarised in a direction about 100° from the sun, and less so at other angles. Although the colours are distorted, the contrast of most clouds against blue sky can be increased by observing through a polarising screen. This is very useful in black and white photography.

The scene, shown normally in *a*, was in Wimbledon at sunset looking towards the southwest in July. In *b* the polar screen was oriented to exclude the blue sky and the exposure was trebled. The sides of the clouds towards the sun are about the same intensity as before but are much the brightest parts of the scene. The blue is almost black.

As the screen is rotated in front of the eye not only does the blue darken but the cloud appears to brighten, as if that light were polarised at right angles to the blue. This is an illusion and is caused by the opening up of the iris of the eye as the blue darkens.

The scene is shown in *c* with the screen arranged to allow the polarised blue light to pass, with the same exposure as *a*. Although the blue is darkened by the general darkening due to the screen, the cloud light is much more darkened.

Polarisation of light is very common in nature. Light reflected from a water surface is very highly polarised, as are

13.5.2 a

13.6.1 a

13.5.2 b

13.6.1 b

13.5.2 c

13.6.1 c

rainbows. Cloud bows are much more easily seen through a polar screen. On the other hand polaroid sunglasses which exclude the glare from water surfaces, have no effect on light reflected from metal, which is a conductor, because it is not polarised.

Many people are able to detect the polarisation of the sky by means of Haidinger's brushes. These are two brown sectors seen radiating in opposite directions from the point at which the eye is directed, with bluer sectors in between, when the light being looked at is polarised. They are best seen at sunset: if one then looks at the zenith the brushes appear in a blue sky. If one turns round continuously at about one rotation in 10 seconds the brushes appear, but fade in about four seconds if the direction relative to the eye is left constant. Some people are unable to see them. The direction to the sun bisects one of the brown quadrants.

# 14 ROTATION

The principal mechanism producing intense rotation is horizontal convergence of air that is already rotating. If the angular momentum is conserved the velocity of a particle around an axis is inversely proportional to its distance from the axis. The most intense effects occur when upward motion causes horizontal convergence in air that is rotating more rapidly than the average to begin with. Upward motion by itself is not enough, and most thermals and convection clouds draw in air horizontally beneath them without any significant rotational effect. When rotation is sufficiently intensified the centrifugal forces produce a low pressure in the centre, in which cloud is often condensed: further convergence can then only occur if the rotation is retarded by friction.

In this chapter we are concerned with the visible manifestations of the effects of rotation in clouds as seen by a single observer.

## 14.1 Dust Devils

**14.1.1** Hot dry surfaces often have a very poor thermal conductivity because of the air spaces between the sand particles. Consequently they become very hot, like tin roofs and car bodies, in sunshine. When cold air passes over them they are rapidly cooled, and because there is no storage of heat beneath the surface which can be readily conducted to the surface, no part of the surface can act as a source of thermals for more than a very short time. Consequently a very unstable temperature gradient can be built up in the lowest layers of air. In this situation a moving vortex can travel over the ground, maintaining a continuous supply of hot air close to the surface where the friction makes the convergence possible, and feeding this air up the central column, thereby maintaining the low pressure.

Dust devils appear most readily on the edge of an area of greater roughness which becomes a line along which the wind strength changes. The convergence then draws vorticity into the whirl rather like the growing whirls in billows. The direction of rotation may be cyclonic or anticyclonic.

The strong wind near the centre raises the dust, and because of the greater friction on the side on which the rotation augments the wind, more convergence occurs there and the whirl moves across the wind towards that side, along a curved track.

Dust devils usually last from a few seconds up to a minute or two, but may last for ten or more minutes if the supply of rotating air is maintained. Often, in the collapsing stage of one which does not have much vertical development, cold air appears in the centre, presumably entering at the top. Those with great vertical development usually die when the inflowing air no longer has rotation and the base of the vortex rapidly rises off the ground and the rotation above ceases because of horizontal divergence.

## 14.2 Cloud Layer Vortices

**14.2.1** When a very flat layer of cloud develops a structure because of the cooling by radiation at the top (see 7.3.1) the sinking motion sometimes produces enough convergence at the top to start a rotation which is visible in the cloud top. These vortices are usually rather feeble, but the low pressure produces a visible hollowing of the cloud top (like the water surface in a bath plug vortex). They are usually seen only in very calm conditions where other disturbances are minimised, and the rotation is usually cyclonic, because, although small, there are not usually any disturbances producing anticyclonic rotation in excess of the earth's cyclonic rotation.

This vortex was seen over the Arctic Sea, but they have been observed over France and Queensland for example, always with cyclonic rotation.

**14.2.2** In the trade winds of the eastern north Atlantic the island of Funchal (Madeira) much of which is above 1,000 m. often reaches through the inversion (see 7.3.1) and blocks the flow. The island is to the right of the bright patch of cloud at the far end of the vortices which dominate the cloud pattern in the near part of the picture. Two vortices with the same direction of rotation are clearly visible. The island is inclined at 45° to the wind and does not shed vortices alternately of opposite rotation, whereas some islands more symmetrically placed in the wind produce vortices of alternate sign like a Karman vortex street.

In this stereo picture from a Mercury satellite at about 160 km. above the earth the three dimensional curvature of the surface is clearly perceptible.

14.1.1

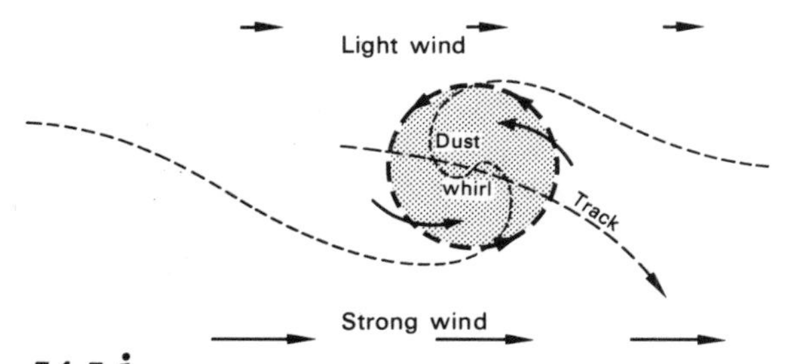

14.1.i

**14.1.i** The track of a dust whirl formed on the boundary of a slower moving body of air is curved towards the faster moving air where the frictional retardation is greater.

14.2.1

14.2.2

## 14.3 Waterspouts and Tornadoes

When violent convection develops in a cloud there is a rapid entrainment of air just below the newly buoyant mass. If this is rotating, instead of ascending up the centre of the thermal it begins to rotate more rapidly and communicates the low pressure to the air below it. Provided the air lower down is also rotating the low pressure is communicated down to the ground or sea surface. On reaching the surface the only significant inflow occurs in the lowest layers where the rotation is reduced by friction. Provided the vortex reaches down to the ground it may last for several minutes, otherwise, if it is filled at the bottom by non-rotating air, the vortex quickly recedes into the cloud.

**14.3.1a** This satellite view, together with the others in this chapter are taken from the first colour films ever recovered from a satellite in orbit. Just east of the Bahamas are seen lines of tall rapidly sprouting cumulus with cirrus floating at the level of the tallest anvils and fields of small cumulus between. At the near end of the anvil of the most westerly line the tall tower has a vortex embedded in it. This is shown in more detail in 14.3.1b in views taken a few seconds earlier.

**14.3.1b** One of several lines of cloud apparently developing rapidly and curving round the centre of a newly growing storm east of the Bahamas. (See 14.3.1a.)

**14.3.2** At the base of a waterspout spray is thrown up like the dust in a dust devil. The funnel cloud extends nearly to the sea surface and has a hollow centre, probably because there is less spray there, through centrifuging. The fine spray can be seen surrounding the cloud. The track of the vortex across the water can be seen leading to the small circular centre beneath the spray. The rotation appears to be anticyclonic for the Mediterranean Sea, which is unusual, but the picture may have been reversed in its original printing.

**14.3.3** Waterspouts are fairly common in waters near coasts, such as near Portugal, in the Mediterranean, etc when the weather is showery. This example shows an oil slick near the coast of N Vietnam which has been wound up by several waterspout whirls. Waterspouts typically occur several at a time or in quick succession, each lasting for a few minutes, and many not reaching down to the sea surface. In this one cloud is formed only in the core of the vortex, although the rotation clearly extends to the surface over a wider area.

Waterspouts like this occur in weather that is otherwise fairly calm at the sea surface, although some are fairly violent like tornadoes—indeed the most violent are tornadoes which have crossed the coast from the land.

**14.3.1 a**

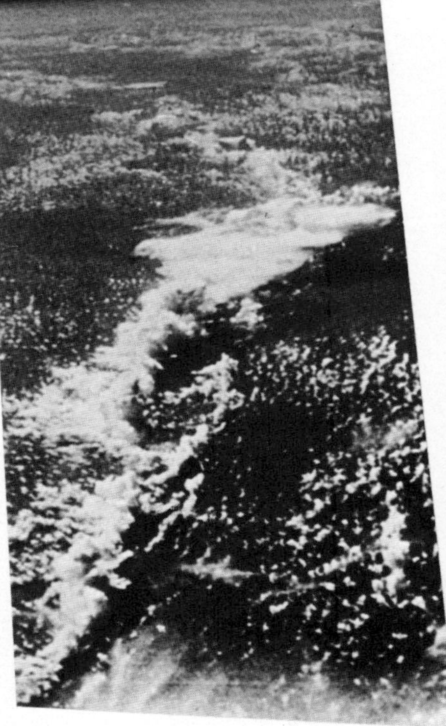

**14.3.1 b**

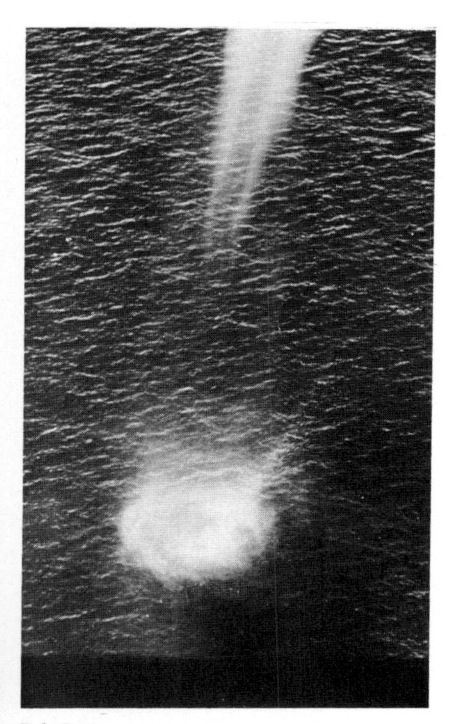

**14.3.2**

**14.3.4**

**14·3·4** Tornadoes always have cyclonic rotation. Occasional funnels on clouds have been seen with anticyclonic rotation but these do not reach the ground because there is never a sufficient supply of air with negative rotation. Cyclonic rotation is common in all storms because they are always accompanied by convergence which intensifies the cyclonic rotation possessed by all large masses of air.

The radar shows only the dense cloud areas and certain fixed objects near to the transmitter. The times are 14 and 20 minutes between successive pictures and the rings are at intervals of 5 nautical miles. The centre of the vortex of a new development, which is on an extension from the main cloud, is gradually covered by a general growth of cloud. It is travelling eastwards at about 20 kt.

**14·3·5** There were five successive tornadoes reaching the ground from this storm near the tip of the warm sector of a cyclone in N. Dakota. Some warm sector air was lifted by the outflowing cold air from intense rain to the west of it on or close to the cold front. The third vortex is shown here; it extended to the ground for 25 minutes and was near the centre of the circular rotating cloud seen from a distance of 7 miles in *c*. There was a series of steps in the base roughly centered on the main vortex with updraughts ranging from 10 to 25 metres per sec. The maximum tangential velocity near the vortex was about 100 m. per sec. near the ground and about half that near cloud base.

The main base of the cloud which was about 14 km. in diameter was at 1,500 metres; the base of the central ring, about $2\frac{1}{2}$ km. in diameter with a base at 350 metres, and with an extension extending in an arc from the centre, called the tail cloud. This lower cloud base occurs where damper air, into which rain has been evaporated, is rising. The tail lay along the edge of the cold outflow of the downdrafts from the most recent rain, and several tails appeared each rotating slowly around the centre. One tail is seen in *a* and *b*.

In *a*, taken 8 minutes after *c*, 5·6 km. from the vortex, the vortex has just reached the ground near the town of Fargo which suffered severe damage. Houses are weakened by the explosive effect of the passage of the low pressure area, where the drop is probably around 50 mb, and torn apart by the strong winds. *b* was taken 10·2 km. away 7 minutes after *a*.

The demise of the funnel occurs when a new one grows and the main convection in the cloud becomes centered there. The new growth usually occurs close to the tail cloud because where two masses of air of different origins meet a vortex sheet is created. When there is a surge of convection above this a vortex can quickly descend to the ground. When the centre of rotation is no longer over a vortex the differences of wind at the top and bottom of it cause it to be sheared over and stretched into what is called a rope cloud. There is no new supply of rotating air at this stage but the rotation is maintained by the stretching. In *d* the rope is seen extending from 340 metres above the trees to 1,700 metres where it enters the cloud: it is leaning over at about 45° and is therefore about 2 km. in length.

The details shown here are taken from Professor Fujita's detailed analysis of all the photographs and other records made available by a multitude of observers in the Fargo area.

**14·3·i** The positions of the lowest part of the cloud and the tail are shown relative to the positions of the observers taking the pictures. In *a*, taken nearly eight minutes after *c*, the tornado has just reached the ground. In *b* it has moved further from the centre, but then rapidly crosses to the other side and is left behind by the cloud which had produced a new funnel 8 km. further on by the time *d* was taken 45 minutes after *c*. The positions of two tail clouds are shown. The times indicated by the numbers are in minutes from *c*, and the dashed line joins the cloud centre to this funnel. The plan position of the rope in *d* is shown.

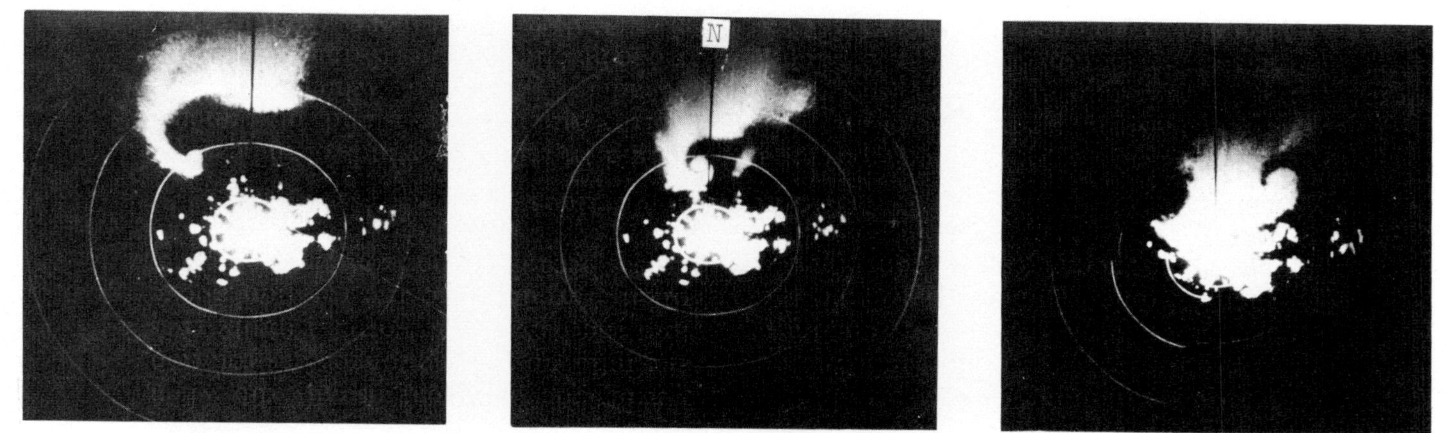


14·3·4 a

14·3·4 b

14·3·4 c

14·3·5 a

14·3·5 b

14·3·5 c

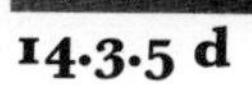

14·3·5 d

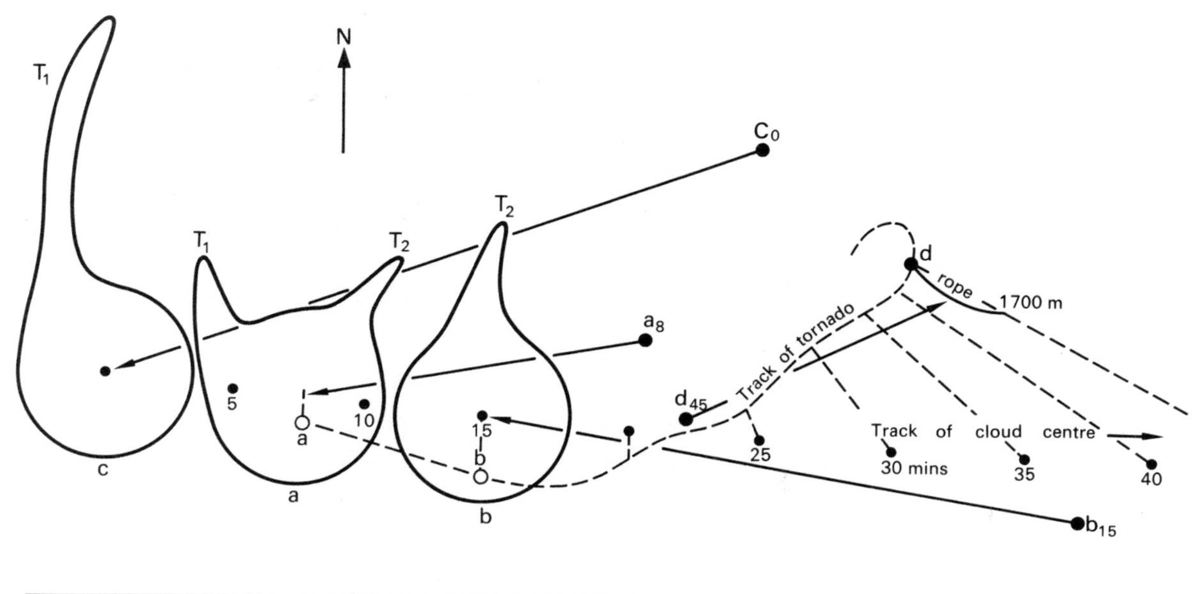


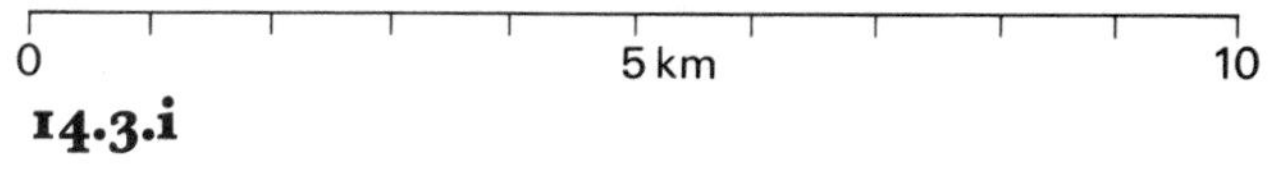


14·3·i

**14.3.6** The ragged cloud around the top of a funnel where it enters the parent cloud is typical of many tornadoes. It indicates that the air in the neighbourhood of the vortex is moister than the average which enters the main cloud base. This lends support to Fujita's theory that they are usually formed at the boundary (front) of the downdraft air as it spreads out on the ground. It is at such a front that the most rapid new growth of cloud above can be expected, and where fairly concentrated vorticity is present below the growth. In this case factory smoke is also sucked in.

There is evidence that many waterspouts which form near coasts grow at boundaries of air masses produced by the flow around headlands at which separation occurs.

**14.3.7** When the main centre of rotation moves from above the point at which the vortex impinges on the ground, the difference in wind speed between the top and bottom of the funnel causes it to be stretched. This funnel is just beginning to be sheared over.

No new buoyancy is being created above the vortex and so its intensity would decrease steadily if there were no stretching. The stretching does make the vortex thinner and soon the cloud is confined to a narrow tube, and is described as a rope cloud. In this picture we can see small patches of the lower cloud beneath the main base, and in the distance the edge of the main cloud. Much of the debris is centrifuged outwards as it is lifted from the small central region of high wind, so that the region containing it widens upwards. Above the main dust cloud particles can be seen to have been centrifuged outwards to a radius of about three cloud tube diameters.

**14.3.8** The situation is very similar to 14.3.7: the main stretching is occurring at about one third of the cloud base height. Not far away is the edge of the storm: this tornado has long passed its peak of ferocity because there are no fragments of low cloud close to it.

**14.3.9** These pictures show the rapid growth of a cloud funnel down to the ground. In strictly laminar motion with uniformly mixed air below cloud base the outline of the funnel would be an isobaric surface at the condensation level. From this we know that in the vortex of a tornado the pressure is commonly 50 mb. less than nearly at the ground.

The rotation reaches the ground before the cloud, and in *b* we see the dust already being raised from the ground. At the same time the sharp tip of the funnel has begun to be eroded.

When the wind at the ground becomes very turbulent there is much mixing and dissipation of energy into heat with the consequence that the air entering the funnel closest to the ground, which is the air which travels most nearly up the centre, is warmed non-adiabatically so that its condensation level is at a lower pressure. Consequently the cloud near the tip of the funnel is formed at a higher level than in adiabatic flow, and in *c* this is clearly seen.

Most of the updraught occurs not at the centre of the funnel but in a ring around it, because the friction is still not enough to allow the air to move on to the axis at the ground. Consequently very odd behaviour is sometimes observed near the axis; the cloud is less dense, and this has been attributed to the centrifuging of cloud particles, which is unlikely because cloud particles are too small to be sufficiently affected in the region of solid rotation at the centre; it has been alleged that there is actually a downdraft there, as is observed in some dust devils. To an aviator in the strong surrounding updraught there would certainly appear to be a downdraft there, and glider pilots have sometimes alleged that their wingtip drops suddenly if, while flying against the rotation in a dust devil, it gets too near the centre of the vortex; this, however, is equally explicable in terms of the horizontal velocity distribution and the lift it would produce on gliders in certain positions.

**14.3.6**

**14.3.7**

**14.3.8**

**14.3.9 a**

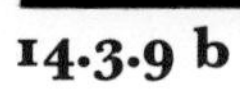

**14.3.9 b**

**14.3.9 c**

## 14.4 Tropical Cyclones

Hurricanes and typhoons occur mostly in the autumn over tropical oceans which are relatively warm (up to around 32°C) at that time. The cyclones are most intense on the western sides of the oceans at the downwind end of the trade wind flow. They sometimes cross the narrow lands of Central America or the Malayan Peninsula and become vigorous on the other side (particularly in the Bay of Bengal).

They travel typically at about 10-15 kts. but may move much more slowly. Although they tend to curve away from the equator this behaviour is not always followed. The winds near the wall of the eye may be 60-100 kts. and the speed usually decreases in proportion to about $r^{1/2}$, where r is the radial distance from the centre of the eye.

The source of energy is in the ascent of air near the centre. This air is made buoyant primarily by the condensation of water vapour into cloud and the deposition of this water as rain. Air near the surface follows a spiral cyclonic track in towards the centre, and as it moves towards lower pressure it becomes cooler, so that convection from the sea surface can be maintained. The main release of heat is by condensation above the cloud base.

The cumulus clouds become arranged by the increasing angular velocity nearer the centre into spiral bands, the cloud towers becoming taller nearer the centre. Layers of cloud are formed where these cumulus spread out, and at the eye wall the ascending cloud forms an almost continuous ring.

The low pressure at the centre is produced by a lowering of the tropopause and a sinking of potentially warm air from the stratosphere down into the eye. The rising air spreads out beneath the tropopause and soon acquires an anticyclonic rotation relative to the earth. It extends a gradually thinning canopy of cirrus outwards from the centre to a radius of 100 km or so.

Some tropical cyclones possess beautiful circular symmetry; others have a relatively inactive area usually in the rear equatorward quadrant.

Tropical cyclones may die as a result of moving over land or cooler sea where their heat and water vapour supply ceases. Or they may be kept going by drawing extratropical fronts into their circulation, and in that case they may continue as vigorous frontal cyclones in the westerlies. They begin in the trade winds where the cumulus are becoming more vigorous (see 3.5.1-3) but the precise mechanism whereby the low pressure centre is first established has been long argued about.

In this section we show typical clouds of tropical cyclones and do not attempt to present a more general discussion of their origin and behaviour.

**14.4.1-3** These stereo pairs were the first to be obtained of a tropical cyclone by a satellite and were not bettered for several years. They show hurricane Debbie whose centre was at 35°N 42°W from a height of about 160 km. The shadow of the cirrus canopy on the sea and low cloud is seen in the upper pictures, and the circular arrangement of the clouds far beyond the active part of the cyclone is seen.

In the middle pictures the thin torrent of cloud down the inside of the hurricane eye, on the right, is evident.

The east side of the storm is shown in the lower pictures and the bands of cirrus can be seen spiralling outwards anticyclonically. As on the west side in the topmost pictures the fallstreaks of cirrus lie radially, showing that the uppermost layers just beneath the tropopause are spreading out most rapidly. Beneath the edges of the cirrus the small cumulus can be seen to be aligned spirally, particularly in the top pictures.

The thin layer of blue on the distant curved horizon is the light scattered from the upper troposphere.

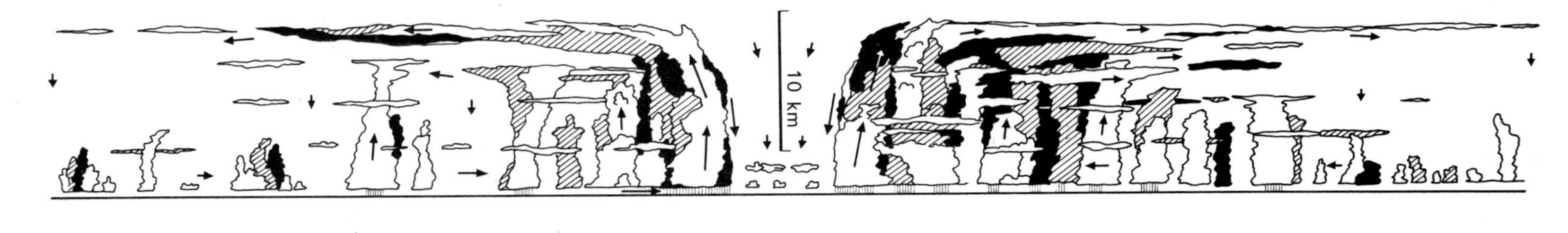


**14.4.i** Section through the clouds of a tropical cyclone.

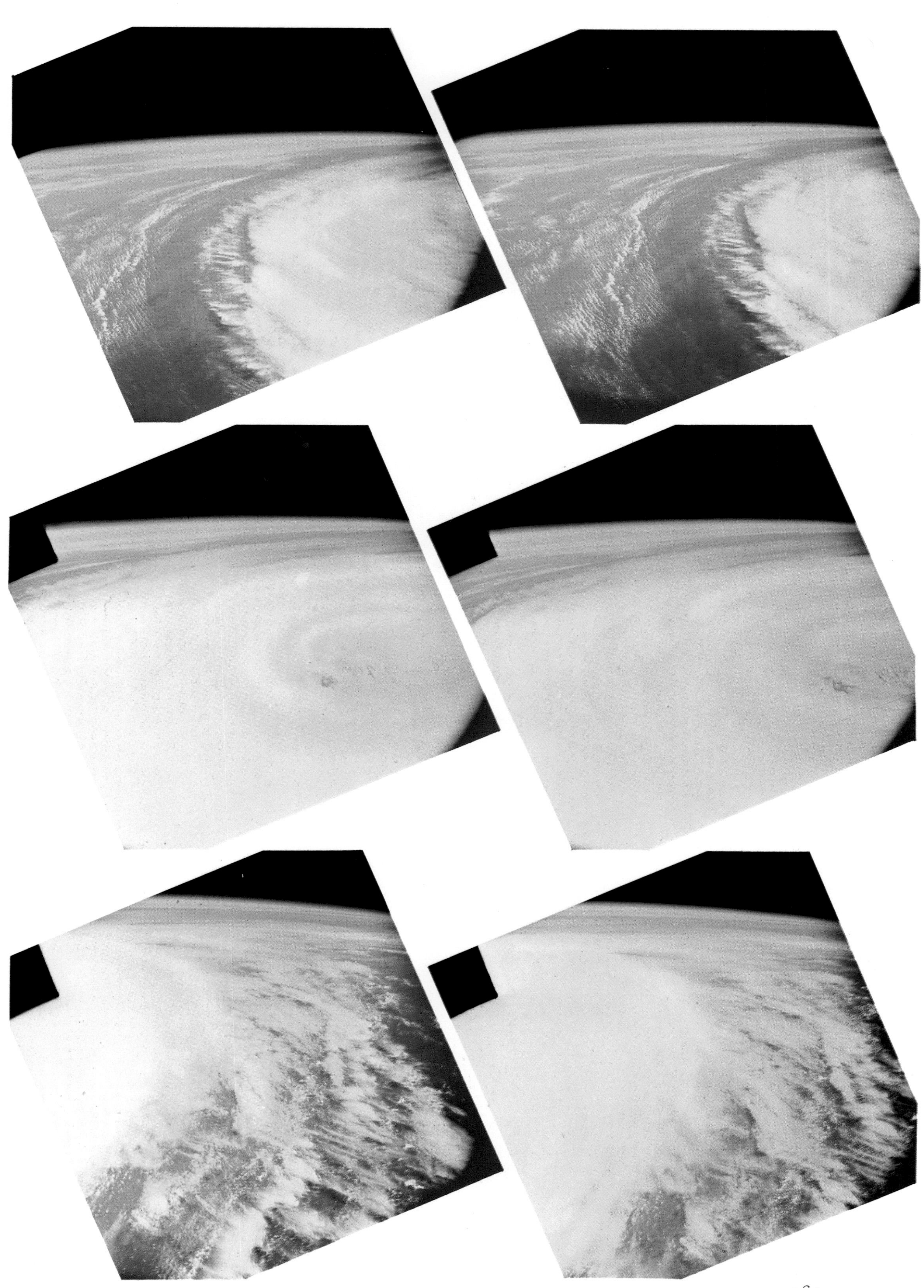

**14·4·4** These two radar pictures of hurricane Donna over Southern Florida show mainly the rain-bearing clouds. The spiral bands are seen in the plan view. Because of the increased speed of rotation nearer the centre any pattern of cumulus clouds would be drawn out into lines of this kind. The track of the air from the edges to the wall of the eye of the storm is similarly spiral, but a particle could make two or more circuits of the eye before beginning to move outwards at higher levels.

The vertical section was made about 8 hours earlier from the same radar station (Key West). The vertical scale is magnified five times. On the right the height of the cumulus towers in the spiral bands increases to a maximum at the eye wall, which slopes so that the eye is about 27 km. wide at the bottom but about 32 km. at a height of 8 km. Within the eye there are clouds up to about 3 km. but these may not be raining.

On the left side, nearest to the radar there is more dense cloud, but the apparently greater height on the extreme left may be because more echo comes from clouds much nearer to the radar set.

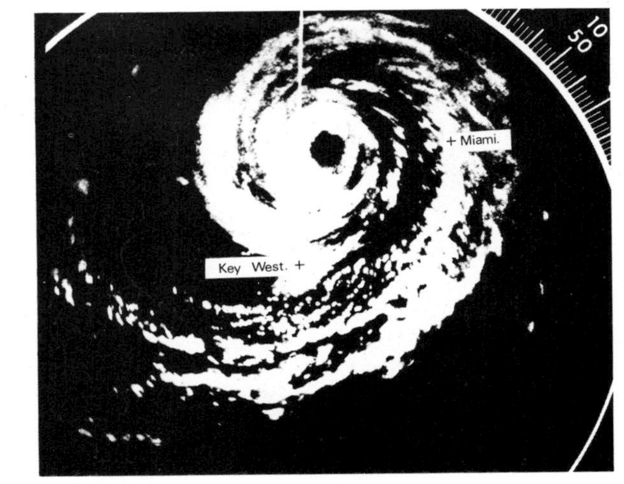


**14·4·4 a**

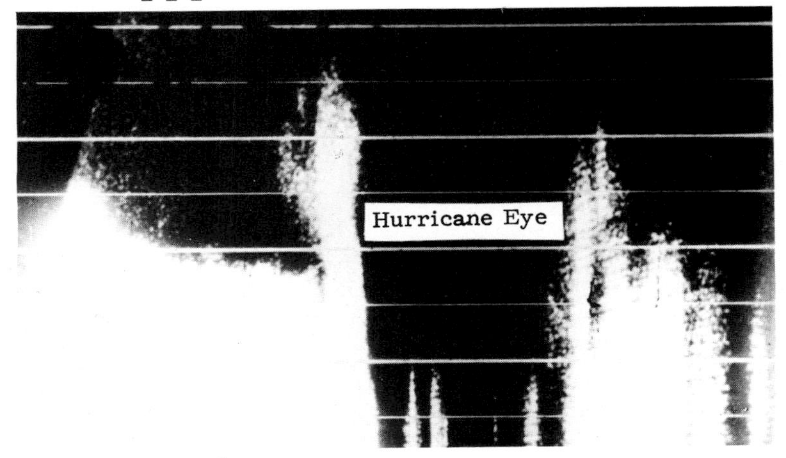


**14·4·4 b**

**14·4·5**

**14·4·5** This view shows how the large wind gradients extend any body of cloud into long spiral lines. The lowest cloud has plenty of gaps through which the sea can be seen but in the distance it becomes both denser and taller, and above the distant towers is the cirrus canopy. The centre of the hurricane is about 110 km. away in the distance and the picture is taken from a height of 5 km.

The most remarkable feature of this scene is the great multitude of layers at which cumulus has spread out and also penetrated. On the left is part of a shower cloud, probably of one of the outer spiral bands.

14·4·6

**14·4·6** The eye of a hurricane is very brightly illuminated in sunshine by the reflection from the wall of cloud surrounding it. The chaotic sea of cloud below may have large gaps through which the sea can be seen, or it may sometimes be a complete cover with a circular arrangement of the cloud (see 14.4.8 and 10).

Sometimes the wall of cloud is almost continuous. The air from which this picture was taken, above the cloud in the eye, was smooth, but in order to travel out of the eye it is necessary first to pass through a region of very strong downcurrent. This is met just before entering the eye wall cloud and extends into it a short distance. The downcurrent is usually turbulent also and it occurs where the cloud is mixing with the very dry stratospheric air which is sinking very slowly down the eye. At the bottom of the eye therefore there is no pure stratospheric air, but only air that has been continuously moistened as it descends near the wall.

The air near the sea in the eye is almost calm, and does not travel as fast as the cyclone; consequently some of it at least is continuously being left behind as the eye moves on, and it is replaced either by this downcurrent air or by air of the rain region being slowly overtaken by the movement of the storm.

The wind speed is very light near the centre, but as the eye wall cloud is approached it rapidly increases and achieves its greatest value just outside the eye.

Cirrus from some of the tallest towers of the wall cloud can be seen towering up to nearly 12 km. above the sea.

14·4·7

**14·4·7** This view is just outside the eye, almost among the most vigorous towers of the wall cloud area. In this case there were some gaps in the wall through which it was possible to fly without entering cloud except momentarily.

Usually the upcurrents are strongest in a ring a few miles wide around the eye and may reach a strength of 30 metres per second. The downcurrent on the inside of the wall cloud may be as strong but is less than half a mile wide usually.

The most noticeable feature of a tropical cyclone is that although it is the largest single convection system in the earth's atmosphere, unlike a frontal cyclone it is really an organisation of very much smaller units which are the individual shower clouds, and which are the most likely form for the instability to take. Between shower clouds, even near the centre of the storm the air may be circulating with very little vertical motion, which is shown by the extensive thin stable layers which some of the cumulus are not buoyant enough to penetrate.

The energy of a hurricane can be measured in a variety of ways. Typically its kinetic energy would be about $5 \times 10^{14}$ kg. $m^2$ $sec^{-2}$, and this is being dissipated continuously by friction at the sea and the creation of eddies by convection currents rising to layers of different velocity (eddy shearing stresses).

If the buoyancy is typically about 1%g (corresponding to an excess temperature of about 3°C) the kinetic energy would require the ascent of 1% of the mass (which is about how much is ascending at any moment) to a height of 400 km. in order for the whole mass to acquire the kinetic energy of a speed of 20 m $sec^{-1}$.

(cont. on p166)

(14.4.7 cont.)
This indicates that the rising air passes through the cyclone much faster than the average and that outside these rising columns there is little dissipation by eddy shearing stresses, for if this were not so, and the air coming in had to generate its own kinetic energy, about 50% of it would have to be rising a distance of around 8 km. But this fraction of the air is not observed to possess the required buoyancy of 1%, and anyway it is not rising, or the space between the lines of shower cloud would be filled with cloud too. This means that the air entering near the bottom is acquiring energy from the pressure field, and the outgoing air is handing it back to the pressure field.

It is erroneous to compute the energy of the hurricane on the basis of the latent heat liberated by the rain, because this heat only becomes kinetic energy of the wind system as the result of the action of buoyancy forces. Thus the sunshine in the Sahara desert produces very little in the way of wind systems because the air is too stable, and this is because it is too dry. Consequently no hurricanes could occur on Mars where there are almost no clouds.

In this picture snow is falling from the overhanging anvils of the larger clouds: below, the occasional castellanus quickly evaporates. In so far as a dramatic theme may be embodied in any natural but inanimate event the tropical cyclone is probably the most suitable. Although it possesses a structure which is unfolded with great force as it passes by or as the observer traverses it by plane, it is an assemblage of individual clouds whose arrangement is peculiar to each hurricane. Its character cannot be hidden by the characters of its elements whereas the frontal cyclone disintegrates upon analysis into air masses, frontal clouds, jet streams, and a multitude of other features each as dramatic as the whole cyclone.

**14.4.8** This colour picture of the eye wall in the distance and the hub at the centre of the eye on the left cannot display the brightness of the scene.

Only the glaciated part of the cumulus tops in the eye wall remain to spread out at the top of the storm as a canopy of cirrus: the rest has descended as rain.

**14.4.10** The centre of the eye reveals its position as the hub of the storm. In this view we see the central two or three miles of the slowly rotating air in the eye. Beneath is the calm region, filled with wretched birds carried by the convergence into the eye, waiting to be swept again into the hurricane as it overtakes them carrying them perhaps once or twice round the eye before depositing them once more in the calm, finally to pitch them ashore breaking them as they try to land for shelter or finally spewing out into a cold ocean those few that remain alive when the circulation dies.

**14.4.12** This 3D view of a typhoon over Japan is included to illustrate the dominance of layer clouds, even in some tropical cyclones. On this occasion there were two typhoons over Japan and this flight began at Osaka, between them, and passed through the edge of the westerly one at this point. In stereo the lines of large cumulus are seen to consist of isolated towers with easily sought out gaps between them.

**14.4.9** We can imagine ourselves to be between the layers of cloud on the right of 14.4.3 looking westwards to the distance in which the cirrus canopy thickens into a continuous sheet and the cumulus grow tall enough to reach up to it.

**14.4.11** This shows an isolated cumulus tower growing at the very edge of a hurricane. It was photographed during an experiment to test techniques for seeding clouds to make them grow and rain. This was a seeded cloud and in these areas it is probable that seeding can be effective on individual clouds, but this is not, unfortunately, a region in which there is any economic value in increasing the rainfall.

Glaciation plays a double role, it decreases the erosion of the cloud tower by reducing or stopping evaporation and it also adds to the buoyancy through the latent heat of freezing. This latter effect will only be important if the surrounding air does not contain many glaciated cloud towers, because only the first few towers grow rapidly as a result of glaciation. They quickly alter the temperature of the environment so that no further spectacular growth can occur.

**14.4.13** Glaciated anvil cloud spreading out from the centre of a small but violent typhoon over Japan: in the foreground are some cumulus tops about 100 km. from the eye of the storm.

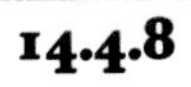

14.4.8

14.4.9

14.4.10

14.4.11

14.4.12

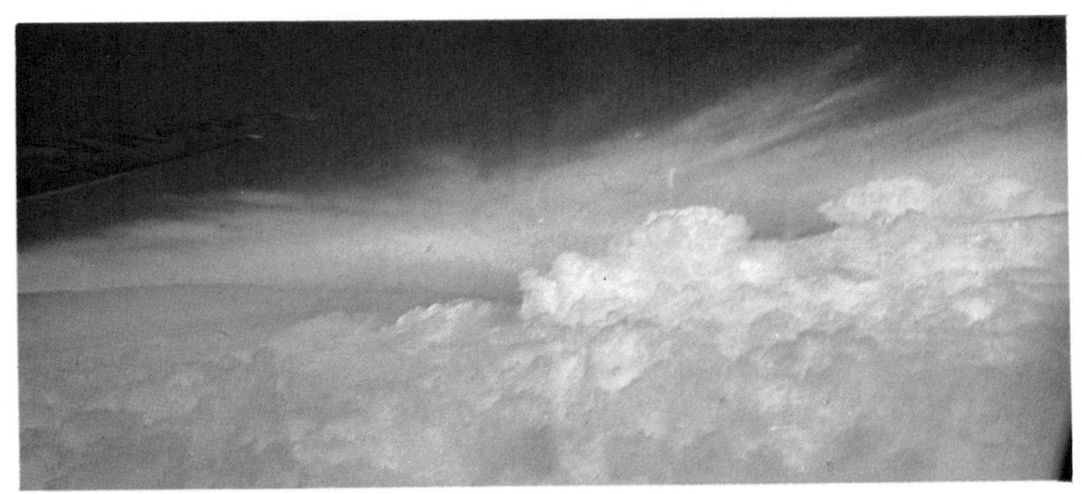

14.4.13

In this book we have merely looked at clouds, seeking to understand them and learning to recognise essential features quickly. In the long run this can only be done through the practice of constant observing: this is not, therefore, a comprehensive text on all that might happen, but an offering in the art of observing.

The phenomena chosen for emphasis have been on the scale most easily seen yet, paradoxically, worst described, indeed almost ignored in standard books on meteorology and cloud physics.

No space has been given to the possibility of artificial control of clouds in any useful way, but it is appropriate to end with tropical cyclones because they are probably most worthy of attention in this connection. This is a very contemporary viewpoint of course, but cloud seeding does not offer a prospect of improving water supplies nearly as promising as desalination of sea water. Tornadoes and hailstorms are the other chief sources of damage which we might seek to inhibit, but they are so localised, swift moving and short lived that it is difficult to lie in wait for them even if we knew what to do with them, which we don't. Major causes of extratropical floods also seem to be too gigantic for us to have any hope of control even if we knew how in theory, which we don't. Tropical cyclones can now be seen near their beginnings by means of satellites often a few days before they do any damage. Thus there could be time for operations to have some effect—if we knew what to do with them, which we don't. But the mere fact that they are organisations of clouds of a kind we can influence individually at the stage when the cyclones begin does offer some hope perhaps, of altering the air on which a new cyclone feeds.

# APPENDIX

## Photogrammetry and Stereoscopic Photography

By means of stereo viewing from a long base line it is possible to obtain a three dimensional perception of clouds. Two quite distinct purposes can be served. The aerial pictures are taken from a base line of very uncertain length, and the camera is not always aimed correctly for measurements to be made on the pictures, but a much clearer view of the relative position of rather thin and often transparent clouds can be obtained. These views can be made use of for the purpose of getting a correct picture of reality before it is represented by a very much simplified model in a theoretical treatment. Perhaps more important is the ability to see a natural scene correctly and quickly, when viewing it as a single observer, that comes from examination of many scenes in stereo.

The second purpose can only be achieved with the aid of optical equipment designed specifically for the evaluation of stereo pictures. By this means positions of cloud features can be measured very accurately, often with surprising results. The two examples shown here are from the work of J. Reuss, who has used very high precision optical systems for the analysis.

A thin layer of altocumulus was photographed simultaneously at the zenith from the ends of a 930 m. base line. The cloud, shown in *b* has a portion enlarged in *a* and the topography of it is shown in plan in *c* and in section in *d*. The heights are in metres above 3,500 m. The cloud patches are thicker in the middle and are concave downwards. The most obvious explanation of the shape is that where the cloud was evaporating it had sunk relative to the thicker parts. It is not known whether this kind of cloud has the same shape at the moment of formation. This cloud evaporated completely within four minutes. Previous measurements on similar clouds had shown the same concave shape. Here the edges were tilted as much as 40° to the horizontal.

Cirrus streaks are particularly suited

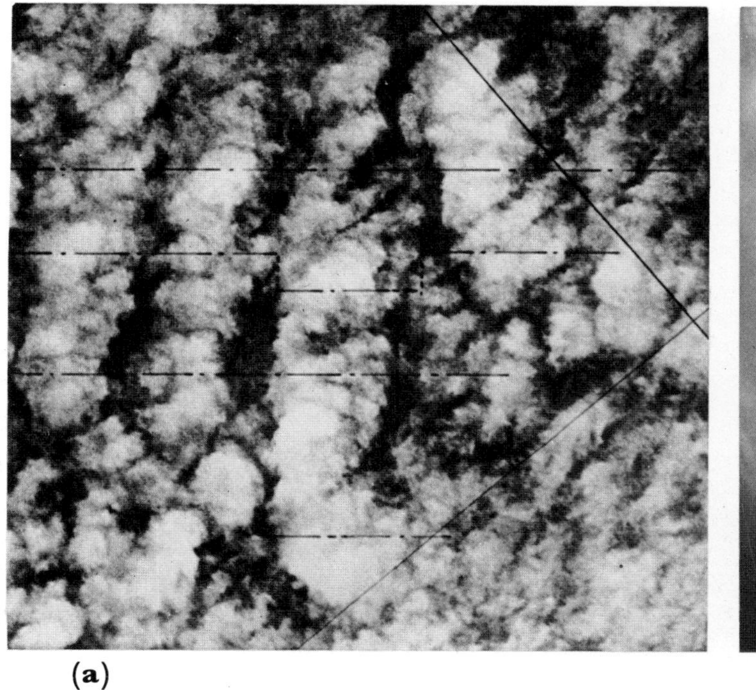

**(a)**

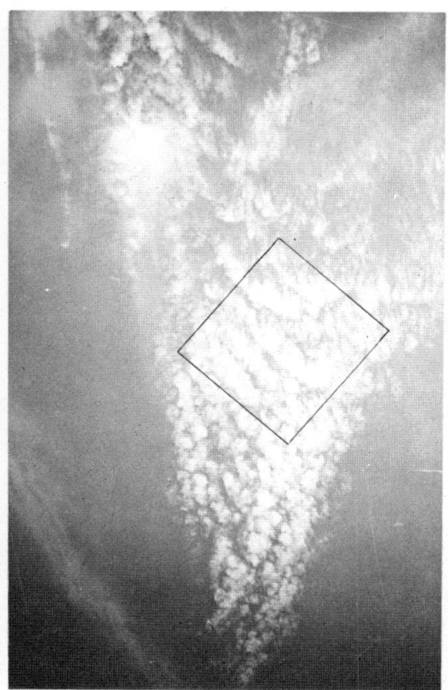

**(b)**

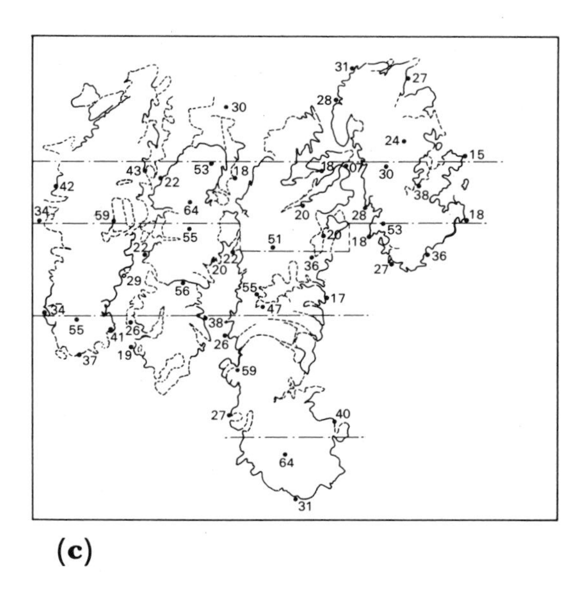


**(c)**

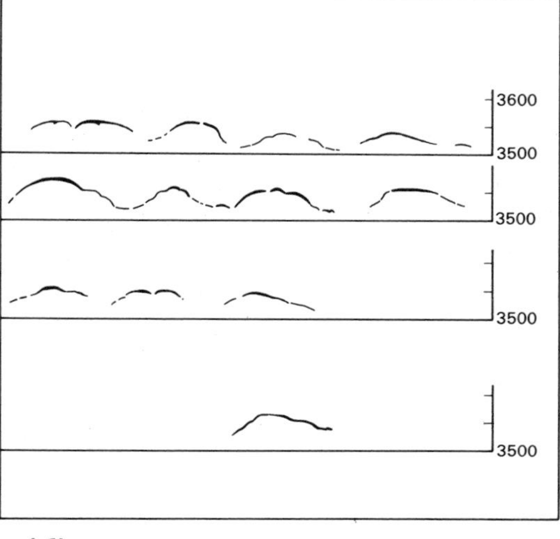


**(d)**

to accurate measurement. This example was taken again with the camera axes perpendicular to the base line but at 45° to the horizontal. The points A, B and C on the picture are shown in the plan and vertical section of the cloud. The 'leaf' is inclined at about 10° to the horizontal. The thin layer of cirrus through which the 'leaf' was seen was measured to be at 5·3 km., which was rather unexpectedly low. The relative height of such clouds is difficult to estimate from the ground because they do not cast appreciable shadows. The streak lay as would be expected if the particles were falling through the wind shear which was measured 23 minutes later (in air 20 km. behind) from the displacements of other cirrus streaks. (Analysis on p 170.)

These diagrams show the plan and elevation of the cirrus streak in the photograph on p.169, as measured by J. Reuss.

The sinking of the particles as they are carried by the wind shear from A towards C is clearly shown. The tenuous nature of this cloud shows that the surrounding air must have been saturated for ice, otherwise the particles would have quickly evaporated.

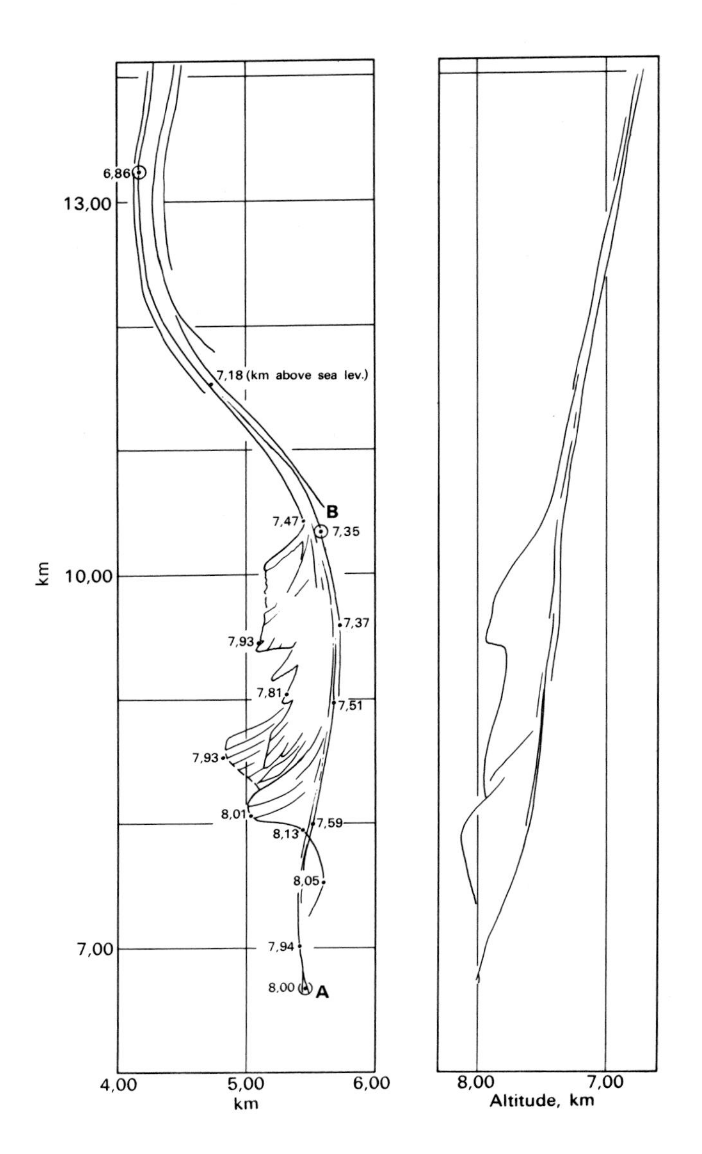


### Taking stereo pictures

From the air two pictures should be taken at an interval of one to three seconds, according to the distance of the clouds. It is helpful to fix attention on a cloud feature and put it in the middle of the two pictures.

When the wind is strong stereo pictures may be obtained from a single point of observation on the ground by allowing the wind to carry the cloud to a different position in the sky. Slightly false results are obtained if the wind shear is strong or the cloud is developing rapidly, but the method is easy to use.

Pictures of clouds in stereo may be obtained by two photographers or by a single photographer in a train, but care should then be taken to avoid including foreground too close to the cameras that is too different in the two pictures. Cliff tops and high viaducts are therefore good viewpoints for this purpose.

### Viewing stereo pictures

The chief difficulty in making use of stereo pictures is in finding an easy method of viewing. The crossed eyes method is superior to any others because it does not require any equipment, but only about one person in three can use it easily and only about half the rest can be bothered to learn. But it is, even then, available to more people than a method requiring lenses, and for this reason the pictures in this book are presented for viewing in this way.

In normally sighted people who do not use glasses or who use the same spectacle lenses all the time, the muscles which control the convergence of the lines of sight of the two eyes and those which bring the object into focus become tied as a result of use. It is therefore difficult to focus on an object at a distance at which the lines do not meet, and ability to do this must first be learned. When it has been learned the viewing of the pictures in stereo is easy, and one's own picture pairs can be displayed on the wall as enlargements for stereo viewing.

Two identical objects may be viewed on a table as one in exaggerated stereo if placed side by side, the right one being viewed with the left eye, the left one with the right eye. At first when the eyes are crossed everything on the table is seen double, but if two of the images are made to coincide so that only three images are seen, the central image is an exaggerated stereo view because the angle between the lines of sight is greater than if a single object had been viewed at the same distance.

The exercise now requires that this exaggerated image should be brought into focus. If they are placed close together, say one inch apart, at first the focusing will be easy if they are viewed from a distance of two or three feet. If the head is brought nearer or the objects moved apart and the still kept as three images, with the middle one double, the focusing becomes more difficult, but can be achieved with practice.

It is important, in doing this, that the lines joining identical parts of the two objects should be parallel to the line through the two eyes, otherwise it is almost impossible to superpose the two inner images of the four, to make only three. This is easily achieved by tilting the head sideways slightly, as necessary.

In order to get to the stage of viewing the pictures, the objects must be seen in this way, in focus, when they are about 3 inches apart. When looking at the pictures the book should be tilted upwards so that the line of sight is more nearly perpendicular to it, and to avoid reflections of lights above one's head in the shiny paper.

When learning to obtain good focusing of the exaggerated stereo image of two identical objects it is often helpful, when the four images have been converted into three, to exclude the distraction of the outer two images by bringing one's hands or two cards in from the sides at a distance about half way between the

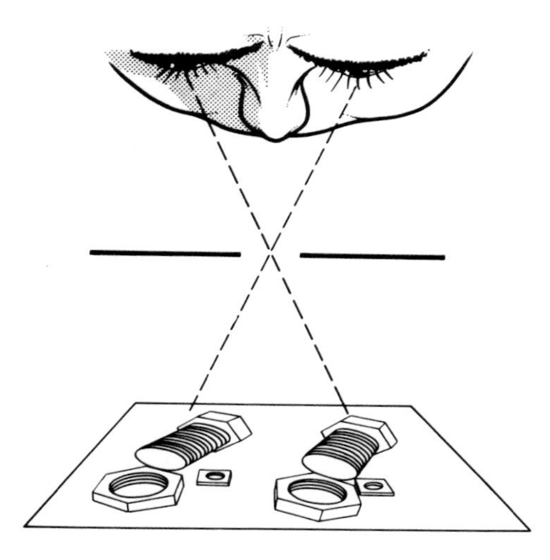

eyes and the objects. By this means, especially if the cards are black, concentration on the exaggerated stereo image is made easier and the focusing muscles work more readily.

With practice the stereo image can be seen within one or two seconds of looking at a picture pair: it is advisable to concentrate the attention at first on obtaining coincidence of a part of the picture where the features stand out with most contrast. The edges of the pictures should be ignored as far as possible, because these cannot be made to coincide unless the same angle of viewing and photographing is used for the two pictures. Also, in some cases it would mean excluding some marginal parts of pictures which are quite interesting.

# BIBLIOGRAPHY


**Historical**

| | | | |
|---|---|---|---|
| 1803 | Luke Howard | On the modification of clouds, and on the principles of their production, suspension, and destruction | J Taylor (London) |
| 1894 | W Clement Ley | Cloudland | Edward Stanford (London) |
| 1905 | Arthur W Clayden | Cloud Studies | John Murray |
| 1920 | G A Clarke | Clouds | C U Press |
| 1926 | C J P Cave | Clouds | C U Press |
| 1926 | W J Humphreys | Fogs and Clouds | Waverley Press (Baltimore) |
| 1943 | Ann Douglas | Cloud reading for pilots | John Murray |
| 1957 | F H Ludlam and R S Scorer | Cloud Study | John Murray |
| 1963 | R S Scorer and H Wexler | A Colour Guide to Clouds | Pergamon Press |
| 1968 | R S Scorer and H Wexler | Cloud Studies in Colour | Pergamon Press |

**Physics, Optics, Pollution**

| | | | |
|---|---|---|---|
| 1920 | W J Humphreys | Physics of the air<br>Later editions 1929, 1940<br>Reprinted 1964 (Optics, part IV) | <br>McGraw Hill (N Y)<br>Dover |
| 1940 | M Minnaert | Light and Colour in the open air | Dover (1954) |
| 1959 | Carl B Boyer | The rainbow: from myth to mathematics | Thomas Yoseloff (London & N Y) |
| 1960 | B J Mason | Physics of Clouds Second Ed 1970 | Clarendon Press (Oxford) |
| 1968 | R S Scorer | Air Pollution | Pergamon Press |
| 1970 | R A R Tricker | Introduction to Meteorological optics | Mills & Boon (London) & Elsevier (N Y) |

**Photographic**

| | | | |
|---|---|---|---|
| 1953 | A Becvar and B Simak | Atlas Horskych Mraku | Prirodovedicke Vydavatelstvi (Prague) |
| 1967 | Yozo Itoh and Shoji Ohta | Cloud Atlas: an artist's view of living cloud | Chijin Shetan (Tokyo) |

**Official Weather Coding**

| | | |
|---|---|---|
| 1956 | International Cloud Atlas | World Meteorological Organisation |
| 1962 | Cloud types for Observers | Meteorological Office (HMSO) |

**Colour Slides and Filmstrips** (Diana Wyllie Ltd, London)

| | | |
|---|---|---|
| 1957 | F H Ludlam and R S Scorer | Clouds (49 frames) |
| 1959 | R S Scorer | Optical Phenomena (10 frames)<br>Storms (10 frames)<br>Stable Weather (15 frames)<br>Unstable Weather (15 frames) |
| 1959 | R S Scorer | Air Pollution (64 frames) |
| 1963 | R S Scorer and H Wexler | A Screen Colour Guide to Clouds (52 frames) |
| 1968 | R S Scorer and J B Andrews | Cloud recognition (15 frames) |
| 1969 | R S Scorer and A B Fraser | Cloud forms (48 frames) |
| 1970 | R S Scorer | The Environment: Air Pollution (79 frames) |

# INDEX

The numbers give individual pictures, sections, or chapters which are relevant.

# CLOUDS OF THE WORLD

1.1.11

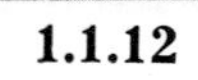

1.1.12

3.3.6

5.1.1

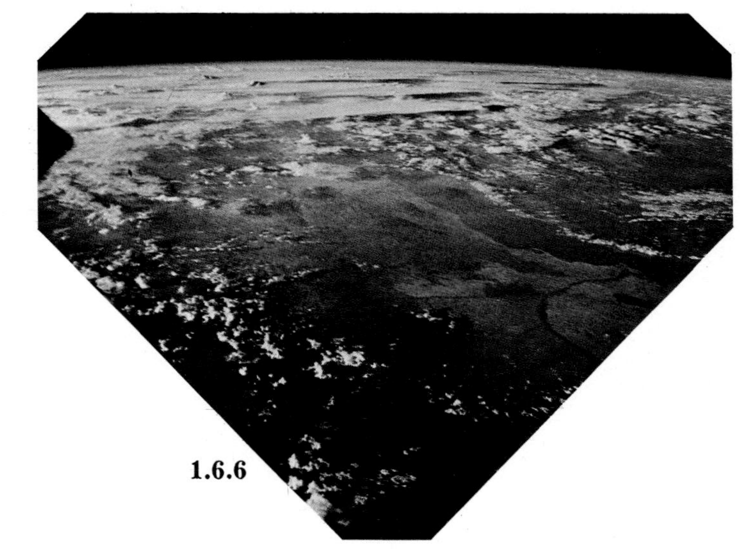

1.6.6

6.2.5

14.4.1

2.3.2

1.4.8

1.6.5

5.1.2

6.2.3

7.1.1

14.4.12

14.3.1 a

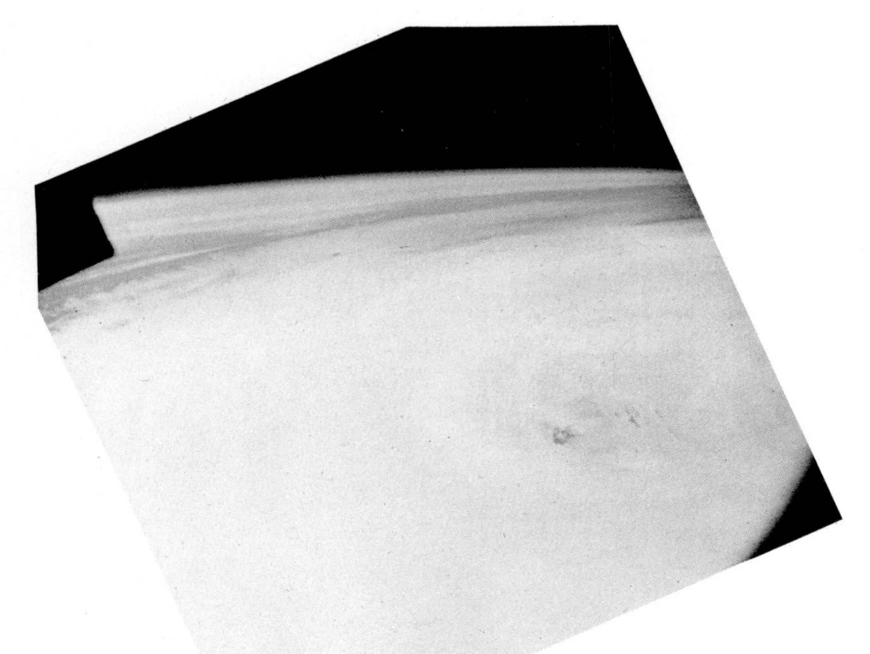
14.4.2